CHAPMAN & HALL/C
Texts in Statistical Scien

Series Editors
C. Chatfield, *University of Bath, UK*
Jim Lindsey, *University of Liège, Belgium*
Martin Tanner, *Northwestern University, USA*
J. Zidek, *University of British Columbia, Canada*

A First Course in Linear Model Theory
Nalini Ravishanker and Dipak K. Dey

Analysis of Failure and Survival Data
Peter Jo Smith

**The Analysis of Time Series —
An Introduction, Fifth Edition**
C. Chatfield

Applied Bayesian Forecasting and Time Series Analysis
A. Pole, M. West and J. Harrison

Applied Nonparametric Statistical Methods, Third Edition
P. Sprent and N.C. Smeeton

Applied Statistics — Principles and Examples
D.R. Cox and E.J. Snell

Bayesian Data Analysis
A. Gelman, J. Carlin, H. Stern and D. Rubin

Beyond ANOVA — Basics of Applied Statistics
R.G. Miller, Jr.

Computer-Aided Multivariate Analysis, Third Edition
A.A. Afifi and V.A. Clark

A Course in Categorical Data Analysis
T. Leonard

A Course in Large Sample Theory
T.S. Ferguson

Data Driven Statistical Methods
P. Sprent

Decision Analysis — A Bayesian Approach
J.Q. Smith

Elementary Applications of Probability Theory, Second Edition
H.C. Tuckwell

Elements of Simulation
B.J.T. Morgan

Epidemiology — Study Design and Data Analysis
M. Woodward

Essential Statistics, Fourth Edition
D.A.G. Rees

Interpreting Data — A First Course in Statistics
A.J.B. Anderson

An Introduction to Generalized Linear Models, Second Edition
A.J. Dobson

Introduction to Multivariate Analysis
C. Chatfield and A.J. Collins

Introduction to Optimization Methods and their Applications in Statistics
B.S. Everitt

Large Sample Methods in Statistics
P.K. Sen and J. da Motta Singer

Markov Chain Monte Carlo — Stochastic Simulation for Bayesian Inference
D. Gamerman

Mathematical Statistics
K. Knight

Modeling and Analysis of Stochastic Systems
V. Kulkarni

Modelling Binary Data
D. Collett

Modelling Survival Data in Medical Research
D. Collett

Multivariate Analysis of Variance and Repeated Measures — A Practical Approach for Behavioural Scientists
D.J. Hand and C.C. Taylor

**Multivariate Statistics —
A Practical Approach**
B. Flury and H. Riedwyl

Practical Data Analysis for Designed Experiments
B.S. Yandell

Practical Longitudinal Data Analysis
D.J. Hand and M. Crowder

Practical Statistics for Medical Research
D.G. Altman

Probability — Methods and Measurement
A. O'Hagan

Problem Solving — A Statistician's Guide, Second Edition
C. Chatfield

Randomization, Bootstrap and Monte Carlo Methods in Biology, Second Edition
B.F.J. Manly

Readings in Decision Analysis
S. French

Sampling Methodologies with Applications
P. Rao

Statistical Analysis of Reliability Data
M.J. Crowder, A.C. Kimber, T.J. Sweeting and R.L. Smith

Statistical Methods for SPC and TQM
D. Bissell

Statistical Methods in Agriculture and Experimental Biology, Second Edition
R. Mead, R.N. Curnow and A.M. Hasted

Statistical Process Control — Theory and Practice, Third Edition
G.B. Wetherill and D.W. Brown

Statistical Theory, Fourth Edition
B.W. Lindgren

Statistics for Accountants, Fourth Edition
S. Letchford

Statistics for Technology — A Course in Applied Statistics, Third Edition
C. Chatfield

Statistics in Engineering — A Practical Approach
A.V. Metcalfe

Statistics in Research and Development, Second Edition
R. Caulcutt

The Theory of Linear Models
B. Jørgensen

ANALYSIS OF FAILURE AND SURVIVAL DATA

Peter J. Smith

CHAPMAN & HALL/CRC

A CRC Press Company
Boca Raton   London   New York   Washington, D.C.

Library of Congress Cataloging-in-Publication Data

Smith, Peter J., 1954–
 Analysis of failure and survival data / Peter J. Smith.
   p. cm. — (Texts in statistical science)
 Includes bibliographical references and index.
 ISBN 1-58488-075-9 (alk. paper)
  1. Regression analysis. 2. Failure time data analysis. 3. Survival analysis (Biometry)
 I. Title. II. Series.
 QA278.8 .S65 2002
 519.5′36—dc21   2002067515
                                                                                                CIP

This book contains information obtained from authentic and highly regarded sources. Reprinted material is quoted with permission, and sources are indicated. A wide variety of references are listed. Reasonable efforts have been made to publish reliable data and information, but the authors and the publisher cannot assume responsibility for the validity of all materials or for the consequences of their use.

Neither this book nor any part may be reproduced or transmitted in any form or by any means, electronic or mechanical, including photocopying, microfilming, and recording, or by any information storage or retrieval system, without prior permission in writing from the publisher.

The consent of CRC Press LLC does not extend to copying for general distribution, for promotion, for creating new works, or for resale. Specific permission must be obtained in writing from CRC Press LLC for such copying.

Direct all inquiries to CRC Press LLC, 2000 N.W. Corporate Blvd., Boca Raton, Florida 33431.

Trademark Notice: Product or corporate names may be trademarks or registered trademarks, and are used only for identification and explanation, without intent to infringe.

Visit the CRC Press Web site at www.crcpress.com

© 2002 by Chapman & Hall/CRC

No claim to original U.S. Government works
International Standard Book Number 1-58488-075-9
Library of Congress Card Number 2002067515
Printed in the United States of America   3 4 5 6 7 8 9 0
Printed on acid-free paper

To Mary

Preface

Lifetime data comes to us in many forms, from diverse areas such as clinical trial studies in medicine and reliability assessments in engineering. Such data may show structural complexity, but it is the presence of censoring which sets its analysis apart from traditional statistical techniques. Censoring causes, not so much missing data, but rather the relaying of partial information for analysis. For example, cancer patients still alive at the termination of a clinical trial have censored lifetimes: they provide incomplete information in that we know that their lifetimes were longer than that observed in the trial. Those that analyse such censored data seek to appropriately incorporate the partial information into the statistical methodology used in the analysis.

In itself, this is not new. The famous proportional hazards methods due to Cox (1972) for censored regression data have widespread software use and applicability. However, the assumptions of proportional hazards may not apply to how certain covariates affect survival time responses. *Analysis of Failure and Survival Data* airs linear regression techniques which have received increasing interest in the research literature. Such techniques may offer a distribution-free alternative to proportional hazards, with the added advantage of being the censored data equivalent of well-understood classical linear regression.

Principally, the Buckley-James Method for censored linear regression has been found to be the best performing linear regression method for censored data for the circumstances when a linear model is appropriate and a Cox Proportional Hazards Model is inappropriate. No current texts discuss the current state of research in this generalisation of least squares to censored data. Much recent new research, referenced to the literature, is presented in this area, making this a work which will be of interest to both students and researchers in the field. Many ideas developed in the earlier chapters of the text are designed to help in understanding the later material: mean conditional lifetimes, graphical and diagnostic plots, Kaplan-Meier attributes. They are taken up in later chapters as tools needed in mastering research techniques for censored regression.

This highlights the didactic nature of research preparation. Students ready to embark on their own research find much of the literature mathematically difficult to absorb at the outset. This text aims to bridge the gap for such students, by placing higher level university coursework at one end of the bridge and direct access to mathematically difficult literature at the other. Many of my students from Melbourne's Key Centre for Statistical Sciences have successfully crossed this bridge. (The university contributors to the Key Centre are, Monash University, RMIT University, La Trobe University and Melbourne University.)

The style of the text has an emphasis on understanding the ideas behind the methodology. Certainly, it is a contemporary understanding with software output from S-PLUS and MINITAB making direct appearance. In the final chapter, S-PLUS code is given for the simple calculation of Buckley-James estimators. However, the emphasis is not on the particularity of which software is used, but rather what may be achieved and understood by its use — what diagnostics are important to action when using software, any software, for particular analyses (for example, QQ-plots for both censored and uncensored data).

Sometimes the style is formal. Always, for easy reference, key terms are highlighted in definitions, key results formalised as theorems and key explanations afforded in simple outline as proofs. It becomes instructive that classical familiar data sets are insightfully analysed (for example, the Stanford Heart Transplant data) to shed light on particular techniques. All data sources are referenced to the literature. Contextual examples are nested within each chapter. A suite of exercises is positioned at the end of each chapter.

I continue to owe a great debt to past and present students and colleagues who have helped shape my attitudes to research. They have assisted me with enthusiasm, support, good humour and patience. Many went further; I would particularly like to thank Jiami Zhang for technical assistance and Darshi Arachchige for S-PLUS coding. To the reviewers and publishers, my sincere thanks.

Peter J. Smith
RMIT University
Melbourne

Contents

1	Survival Distributions	1
2	Hazard Models	19
3	Reliability of Systems	37
4	Data Plots	55
5	Censoring and Lifetables	73
6	The Product-Limit Estimator	95
7	Parametric Survival Models under Censoring	119
8	Fitting Parametric Regression Models	143
9	Cox Proportional Hazards	167
10	Linear Regression with Censored Data	187
11	Buckley-James Diagnostics and Applications	215
	References	246
	Index	252

CHAPTER 1
Survival Distributions

1.1 Basic concepts

Survival analysis involves the study of lifetime distributions. By this we mean lifetimes of people, of cancer patients, of industrial robots, of components, of cogs, of software. We consider a broad range of applications both industrial and biological.

▬ Example 1.1
Failure data for reduction cells in aluminium smelting
Lifetime data commonly occur in the engineering environment. In this example from a Canadian aluminium smelter, alumina is liquefied in a steel-lined box (or cell) which is built to withstand extremely high temperatures. In the smelting process, aluminium is produced as a byproduct when the cell functions like a battery with molten alumina as the electrolyte. The cell needs to be replaced when the carbon lining cracks, allowing impurities into the process. The failure time data listed were part of a larger data set and represent days of service until cell replacement.

Failure age (days)					
1540	1415	660	999	1193	1006
869	1035	797	296	775	1424
1169	1500	728	670	841	

(*Source*: Whitmore, G.A., Crowder, M.J. and Lawless, J. (1998). Failure inference from a marker process based on a bivariate Weiner model. *Lifetime Data Analysis*, 4, 229–51.)

□

We will be concerned with not only observing lifetime data, recording it, displaying it, but also consideration of factors affecting lifetimes, explanatory variables that help explain observed lifetimes. A typical example of survival data is given in Example 1.2, where lifetime refers to time from carcinogen injection to time of death in experimental laboratory mice. The data therefore represent death times as well as lifetimes. For this particular data there are observations on both mice kept under standard laboratory conditions and mice kept in a germ-free environment during the course of the trial — these experimental conditions are outcomes of an explanatory variable.

■ Example 1.2
Times to tumor onset in mice

The data which follow were from a study designed to check whether changing experimental conditions has any effect on the 'time until tumor onset' in mice injected with carcinogen. The time of injection with carcinogen was taken as time zero. The mice were then observed: of interest was the length of time until tumors develop in the mice.

Necropsy findings	Individual ages at death (days)							
Conventional mice								
Tumor	381	477	485	515	539	563	565	582	603
	616	624	650	651	656	659	672	679	698
	702	709	723	731	775	779	795	811	838
No tumor	45	198	215	217	257	262	266	371	431
	447	454	459	475	479	484	500	502	503
	505	508	516	531	541	553	556	570	572
	575	577	585	588	594	600	601	608	614
	616	632	632	638	642	642	642	644	644
	647	647	653	659	660	662	663	667	667
	673	673	677	689	693	718	720	721	728
	760	762	773	777	815	886			
Germ-free mice								
Tumor	546	609	692	692	710	752	753	781	782
	789	808	810	814	842	846	851	871	873
	876	888	888	890	894	896	911	913	914
	914	916	921	921	926	936	945	1008	
No tumor	412	524	647	648	695	785	814	817	851
	880	913	942	986					

(*Source*: Hoel, D.G. and Walburg, H.E. Jr. (1972). Statistical analysis of survival experiments. *Journal of the National Cancer Institute*, 49, 361-2.)

Since it is difficult to detect when a tumor develops, mice were inspected for tumors at time of sacrifice or time of death. For these survival data, a 'lifetime' refers to 'time to tumor onset'!

Such a lifetime is hard to measure exactly. Some mice at death or sacrifice showed no evidence of tumors present and this has a complicating effect on the analysis of these data. If, at sacrifice, no tumor is present, then clearly the time to onset would be longer than the sacrifice time. If, at death, a tumor is present, then the time to onset of the tumor is shorter than the time to

SURVIVAL FUNCTIONS

death (since the tumors are not rapidly lethal). These effects are examples of *censoring* and are a common feature of lifetime data. We take up a full discussion of censoring in Chapter 5. □

Finally there is the aspect of prediction: how long will a battery last? How can we construct a guarantee period? What length of remission can we expect for a cancer patient? How long (see Example 1.1) before industrial containers fail? Our ability to predict is often dependent on the analysis of lifetime data. In many contexts (such as in medicine) this analysis is often performed in a distribution-free way. However, in many engineering applications, the analysis may depend on a model which is *fitted* to the data. The accuracy of our inferences is dependent on the goodness of fit of the model. In the course of this text we will be examining both distribution-free and model-based approaches.

1.2 Survival functions

In the language of survival analysis there are many words and phrases that mean the same thing. We clarify the terminology from engineering and medical areas.

By a **lifetime**, or **survival time**, we mean a period from the start of observation until **death** occurs. Death is also termed **failure**, so that the time of death is a **failure time**. Essentially, living objects experience death, whereas non-living objects experience failure.

Definition 1.1 *A random variable Y is a* **survival random variable** *if an observed outcome y of Y lies in the interval $[0, \infty)$.* □

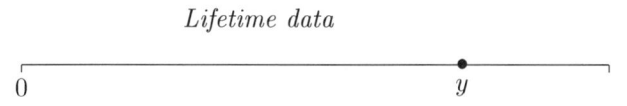

Lifetime data

```
|────────────────────────●────────────────────|
0                        y
```

Collectively, survival data $y_1, y_2, y_3, \ldots, y_n$ are observations of a positive-valued random variable Y. In this book Y will generally be taken as absolutely continuous. It will be clear from the context when Y is to be taken as a discrete random variable.

Suppose that Y has probability density function f and cumulative distribution function F. Then

$$F(y) = P(Y \leq y) = \int_0^y f(u)du. \tag{1.1}$$

Definition 1.2 *The* **survival function** *or* **reliability function**, *S, is defined for all values of y by $S(y) = 1 - F(y)$.* □

This means that

$$S(y) = P(Y > y) = \int_y^\infty f(u)du, \tag{1.2}$$

so that by the Fundamental Theorem of Algebra,

$$f(u) = -\frac{d}{du}S(u). \tag{1.3}$$

The study of survival functions is at the heart of survival analysis. From (1.2) we can establish the following simply because probability density functions integrate to 1:

$$S(0) = \int_0^\infty f(u)du = 1.$$

Further,

$$S(\infty) = \lim_{y \to \infty} S(y) = \lim_{y \to \infty} \int_y^\infty f(u)du = 0.$$

Finally, if $a \geq b$, then

$$S(b) - S(a) = \int_b^a f(u)du \geq 0.$$

This establishes the following theorem.

Theorem 1.1 *The survival function S is monotone decreasing over its* **support** $[0, \infty)$. *Further, S satisfies $S(0) = 1$, $S(\infty) = 0$.* □

In fact, any monotone decreasing function S with support $[0, \infty)$ with $S(0) = 1$ and $S(\infty) = 0$ is the survival function of some survival random variable. The required random variable is the one having probability density function $f(u) = -\frac{d}{du}S(u)$.

The general shape of a survival function for a continuous random variable is illustrated in Figure 1.1.

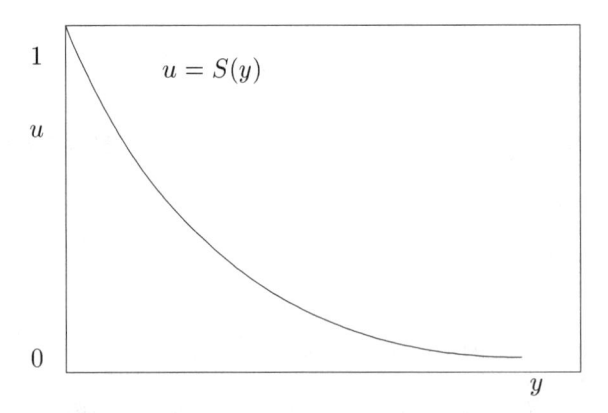

Figure 1.1 *Diagram showing the shape of the graph of the survival function, S, for a survival random variable Y.*

1.3 Using data to estimate the survival function

Suppose that we have obtained data on the survival random variable Y. How can we use the data to estimate $S(y)$? We start by not assuming a model for the distribution for Y, but rather listen to what the data suggest for the shape of the survival function. The simplest (and most obvious) method is through the *empirical survivor function* which effectively counts up the number of data points larger than y.

In this process of counting, we will use the concept of the **indicator function**, I_A, of an event A, defined by

$$I_A(x) = \begin{cases} 1 & \text{if } x \in A \\ 0 & \text{otherwise.} \end{cases}$$

✷ **Definition 1.3** *Given n observations $Y_1, Y_2, Y_3, \ldots, Y_n$ independently and identically distributed (i.i.d.) with the same distribution as Y, the **empirical survivor function** S_n is defined for all values of y by*

$$S_n(y) = \frac{\text{number of observations} > y}{n} = \frac{1}{n}\sum_{i=1}^{n} I_{(y,\infty)}(Y_i) \qquad (1.4)$$

and is an estimate of the survival function $S(y) = P(Y > y)$. □

We show how to calculate $S_n(y)$ from data in Example 1.3.

Definition 1.3 shows that for a fixed value of y, $S_n(y)$ is an average of n independent and identically distributed random variables, Z_i, say, where $Z_i = I_{(y,\infty)}(Y_i)$. Therefore

$$nS_n(y) = \sum_{i=1}^{n} Z_i.$$

The Z_i each have the same Bernoulli probability distribution (that is, are i.i.d.), where $P(Z_i = 1) = P(Y_i > y) = S(y)$ and $P(Z_i = 0) = P(Y_i \leq y) = 1 - S(y)$. This common Bernoulli distribution therefore has probability mass function

z	0	1
$P(Z = z)$	$1 - S(y)$	$S(y)$

Note that this probability distribution has mean $E(Z_i) = S(y)$ and variance $Var(Z_i) = S(y)[1-S(y)]$. This means that $nS_n(y)$ has a binomial distribution: $nS_n(y) \sim Bin(n, S(y))$. In particular, of use for estimating survival at a fixed value of y,

$$E[S_n(y)] = \frac{1}{n}[nS(y)] = S(y) \qquad (1.5)$$

and

$$Var[S_n(y)] = \frac{1}{n^2}\{nS(y)[1-S(y)]\} = \frac{S(y)[1-S(y)]}{n}. \qquad (1.6)$$

At a fixed point $y = y^*$ we estimate $S(y^*) = P(Y > y^*)$, the probability of survival beyond y^*, using the estimator $S_n(y^*)$. Note that (1.5) shows that $S_n(y^*)$ is unbiased for $S(y^*)$. To get a measure of the accuracy by which $S_n(y^*)$ estimates $S(y^*)$, it is appropriate to use an approximate confidence interval based on two standard errors:

$$S_n(y^*) \pm 2\sqrt{\frac{S_n(y^*)[1 - S_n(y^*)]}{n}}$$

Again, for the variance term under the square root sign, S_n is used to estimate S. This calculation is demonstrated in Example 1.3.

■ Example 1.3
Empirical survivor function for the placebo effect

Gehan (1965), Lawless (1982) and others have discussed data from a clinical trial examining steroid-induced remission times (weeks) for leukemia patients. One group of 21 patients were given 6-mercaptopurine (6-MP); a second group of 21 patients were given a placebo — the data $y_1, y_2, \ldots, y_{21}$, which follow in the table, are remission times from this placebo group.

Placebo group
Steroid induced remission times (weeks)

1	1	2	2	3	4	4	5
5	8	8	8	8	11	11	12
12	15	17	22	23			

(*Source*: Gehan, E.A. (1965). A generalised Wilcoxon test for comparing arbitrarily singly censored samples. *Biometrika*, 52, 203–23.)

We now illustrate how to estimate the probability of more than a three-month remission for patients in the placebo group.

Using $S_{21}(12)$ for estimation of $S(12)$, the true probability of survival beyond 12 weeks, we find that four observations lie beyond 12. This means that four of the indicator functions $I_{(12,\infty)}(y_i)$, for $i = 1, 2, 3, \ldots, 21$ have value 1 and the remaining 17 have value 0. Thus,

$$S_{21}(12) = \frac{1}{21} \sum_{i=1}^{21} I_{(12,\infty)}(y_i) = \frac{4}{21}, \quad (1.7)$$

and an approximate two standard error interval for the true probability is

$$S_{21}(12) \pm 2\sqrt{\frac{S_{21}(12)[1 - S_{21}(12)]}{21}} \quad \text{or} \quad \frac{4}{21} \pm 2\sqrt{\frac{\frac{4}{21}(1 - \frac{4}{21})}{21}}.$$

This reduces to 0.19 ± 0.171. The apparent lack of accuracy is caused by the relatively small sample size. □

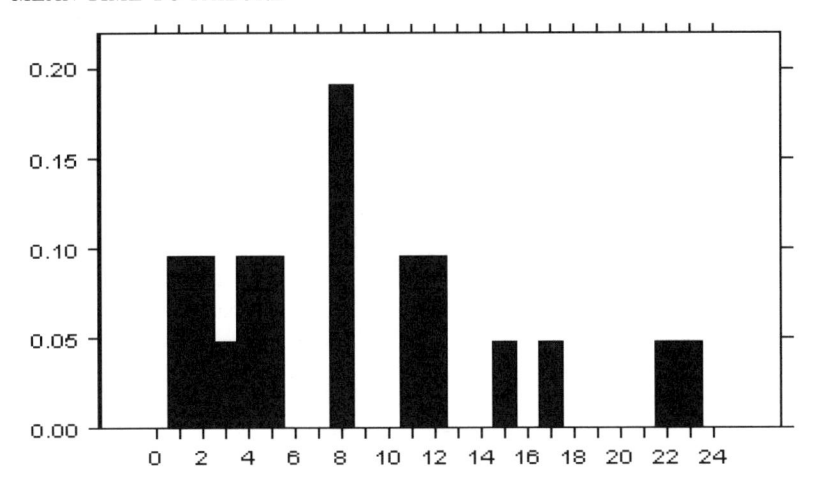

Figure 1.2 *Histogram for the steroid remission data of Example 1.3 where the remission times in weeks lie on the horizontal axis and relative frequency on the vertical axis.*

We can examine the calculation of the empirical survivor function at 12 weeks for the leukemia patients in Example 1.3 and seek a geometric interpretation. The calculation which appears in (1.7) shows that $S_{21}(12)$ is found by assigning $\frac{1}{21}$ to each data point, then adding over all data points larger than 12. If we constructed a histogram for these data, with boxes of width 1 unit standing centred at each data point, then $S_{21}(12)$ would represent the total area of the boxes beyond 12 weeks; that is, the area under the histogram beyond 12. If we examine Figure 1.2, there are four boxes (each of approximate height 0.05 and width 1) lying beyond 12, giving an approximate area of 0.2.

Using calculus, we may view summations as integrals and write

$$S_{21}(12) = \sum_{i=1}^{21} I_{(12,\infty)}(y_i)\frac{1}{21} = -\int_{12}^{\infty} dS_{21}(y)$$

as an estimate of

$$-\int_{12}^{\infty} dS(y) = S(12).$$

1.4 Mean time to failure

Theorem 1.1 indicates that it is not possible to compare survival random variables on the basis of a comparison of their survival functions, since the graphs of two different survival functions may cross. That is, if two survival functions S_1 and S_2 satisfy $S_1(y) > S_2(y)$ for some y, then it does not follow that $S_1(y) > S_2(y)$ for all y. This is illustrated for two survival functions in Example 1.4.

■ Example 1.4
Different types of survival functions
The functions S_1 and S_2 defined on $[0, \infty)$ by

$$S_1(y) = e^{-\frac{y}{2}} \quad \text{and} \quad S_2(y) = e^{-\frac{y^2}{4}}$$

satisfy the requisite properties of survival functions: for each $i = 1, 2$, clearly $S_i(0) = 1$ and $S_i(y) \to 0$ as $y \to \infty$; each S_i is monotone decreasing since $\frac{d}{dy} S_i(y) < 0$. Finally, a comparison of the two survival curves shows that

$$\begin{aligned} S_1(y) > S_2(y) &\iff -\tfrac{y}{2} > -\tfrac{y^2}{4} \\ &\iff 2y < y^2 \\ &\iff y < 0 \text{ or } y > 2 \\ &\iff y > 2. \end{aligned}$$

Therefore, between 0 and 2, S_2 is larger than S_1, whereas, for values of y larger than 2, S_1 is larger than S_2.
□

One way of characterising survival random variables 'on average' is through their means.

Definition 1.4 *If survival random variable Y has expected value $E(Y)$, then the **mean time to failure**, or **MTTF** of Y, is given by* MTTF $= E(Y)$. □

This definition shows that we may write MTTF as the integral

$$\text{MTTF} = \int_0^\infty u f(u) du,$$

and applying integration by parts to the integral on the right gives the mean time to failure in terms of the survival function as summarised in Theorem 1.2.

Theorem 1.2 $\quad \text{MTTF} = \int_0^\infty S(u) du.$ □

Therefore, to find an average lifespan, we integrate the survival function.

■ Example 1.5
Exponential survival functions
If β is a constant, then the function S defined by

$$S(y) = e^{-\frac{y}{\beta}} \quad \text{for } y \geq 0$$

represents the survival function for some random variable Y. (Recall the characteristics of a survival function in Theorem 1.1.) By integration, we find that

$$\text{MTTF} = \int_0^\infty S(u) du = \int_0^\infty e^{-\frac{u}{\beta}} du = \beta.$$

For two such survival functions $S_1(y) = e^{-\frac{y}{\beta_1}}$ and $S_2(y) = e^{-\frac{y}{\beta_2}}$, representing random variables with respective mean times to failure MTTF_1 and MTTF_2, it follows that

$$\begin{aligned} S_1(y) > S_2(y) &\iff e^{-\frac{y}{\beta_1}} > e^{-\frac{y}{\beta_2}} \\ &\iff -\frac{y}{\beta_1} > -\frac{y}{\beta_2} \\ &\iff \beta_1 > \beta_2 \\ &\iff \text{MTTF}_1 > \text{MTTF}_2. \end{aligned}$$

□

1.5 Toward the hazard function

The reliability function examines the chance that breakdowns (of people, of experimental units, of computer systems, ...) occur beyond a given point in time. To monitor the lifetime of a component across the support of the lifetime distribution, the **hazard function** is used.

The probability that a system will fail now, that is, in $(y, y + \Delta y)$, **given that it has lasted up to now**, is given by

$$P(y < Y < y + \Delta y | Y > y) = \frac{P(y < Y < y + \Delta y, Y > y)}{P(Y > y)} = \frac{P(y < Y < y + \Delta y)}{P(Y > y)}.$$

Averaging this over the length of the current time interval Δy gives the **average failure rate** across the interval:

$$\frac{P(y < Y < y + \Delta y)}{(\Delta y) P(Y > y)}. \tag{1.8}$$

In the limit, as Δy tends to zero, the average failure rate becomes an instantaneous failure rate

$$\lim_{\Delta y \to 0} \frac{P(y < Y < y + \Delta y)}{(\Delta y) P(Y > y)} = \lim_{\Delta y \to 0} \frac{F(y + \Delta y) - F(y)}{(\Delta y) S(y)} = \frac{F'(y)}{S(y)} = \frac{f(y)}{S(y)}.$$

This motivates the following definition.

Definition 1.5 *A survival random variable Y has* **hazard function**, *or* **hazard rate** *or* **force of mortality**, *h, defined for $y > 0$ by*

$$h(y) = \frac{f(y)}{S(y)}.$$

□

This is the instantaneous rate of death (or failure) at time y, given that the individual survives up to time y. Hazard functions track how the failure rate changes with time. The survival function is expressed in terms of the hazard function by the following theorem.

Theorem 1.3 $\quad S(y) = e^{-\int_0^y h(u) du}$.

Proof: Since $f(u) = -S'(u)$, it follows that $h(u) = -\frac{d}{du}\log_e S(u)$, and integrating each side of this equation between 0 and y gives

$$\log_e S(u)\big|_0^y = -\int_0^y h(u)du.$$

The result follows from this since $S(0) = 1$. $\qquad\square$

1.6 Life expectancy at age t

Suppose that we have tracked the lifetime of our patient, escalator cog, computer software, ..., to a particular age t. That is, our experimental unit is still alive and under observation at time t. The question of this section is: How much life does it have left in it on average? If survival time is denoted as usual by the random variable Y, and we know that the constant t is such that $Y > t$, then the discrepancy $Y - t$ is called the **residual lifetime at age** t.

Definition 1.6 *For $t > 0$, the* **mean residual lifetime at age** t, $r(t)$, *is given by* $r(t) = E(Y - t | Y > t)$. $\qquad\square$

Noting that t is constant, if we evaluate the expected value in Definition 1.6, we obtain

$$\begin{aligned}
r(t) &= E(Y - t | Y > t) \\
&= \int_t^\infty (y-t) dP(Y \le y | Y > t) \\
&= \int_t^\infty (y-t) d\left(\frac{S(t) - S(y)}{S(t)}\right) \\
&= \int_t^\infty (y-t)\left(\frac{-S'(y)dy}{S(t)}\right) \\
&= \frac{-(y-t)S(y)\big|_t^\infty + \int_t^\infty S(y)dy}{S(t)} \\
&= \frac{\int_t^\infty S(y)dy}{S(t)}.
\end{aligned}$$

In this derivation we have used the result that $yS(y) \to 0$ as $y \to \infty$. (Can you see where?) This result is established in the exercises at the end of the chapter.

We have found that the mean residual lifetime has the convenient integral expression

$$r(t) = \frac{\int_t^\infty S(y)dy}{S(t)}. \tag{1.9}$$

If we evaluate this expression at $t = 0$, we obtain, using Theorem 1.2,

$$r(0) = \frac{\int_0^\infty S(y)dy}{S(0)} = \int_0^\infty S(y)dy = E(Y) = \text{MTTF}. \tag{1.10}$$

Life expectancy at a given age t is a measure of duration of lifetime **given that** age t has already been reached. Formally:

LIFE EXPECTANCY AT AGE T

Definition 1.7 *For a survival random variable Y and a fixed time t, the* **mean life expectancy at age t** *is given by $E(Y|Y > t)$.* □

By (1.9), mean life expectancy at age t is related to residual lifetime by adding in the time t already lived. This means that

$$E(Y|Y > t) = t + r(t) = t + \frac{\int_t^\infty S(y)dy}{S(t)}. \qquad (1.11)$$

■ Example 1.6
Comparing human life expectancy with that of a lightbulb

The left-side data table showing the life expectancy at age t for men and women and various values of t was reported in *The Age*, Melbourne, 4 March 1991, from information supplied by the Australian Bureau of Statistics.

	Life expectancy			Life expectancy
Age now t	Male (years)	Female (years)	Age now t	Lightbulb
0	73.3	79.5	0	73.3
10	74.2	80.3	10	83.3
20	74.7	80.5	20	93.3
30	75.2	80.7	30	103.3
40	75.9	81.0	40	113.3
50	76.7	81.6	50	123.3
60	78.3	82.8	60	133.3
70	81.5	84.8	70	143.3
80	86.5	88.4	80	153.3
90	93.5	94.1	90	163.3

These data show that the survival distributions for men and women are slightly different. When we are born, the expected value of the length of life variable, X, is 73.3 years for men and 79.5 years for women. Given that we live to age t (the left-side column of the table), the future life expectancy beyond t reduces. That is, a man's life expectancy at age 10 is only $74.2 - 10 = 64.2$ more years. This is less than the average lifetime 73.3 years from age 0. It keeps declining: the life expectancy at age 20 is $74.7 - 20 = 54.7$ years.

On the other hand, the life expectancies in the right-side table are for a theoretical lightbulb with survival time random variable $X \sim exp(73.3)$. When this lightbulb is turned on, it has an expected lifetime of 73.3 units of time (and such units will obviously not be years). However, since the lifetime distribution is exponential, it is, in particular, memoryless. That is, the life expectancy of the lightbulb at age 10 units is $83.3 - 10 = 73.3$ units more — still! It does not wear out (unlike human beings, who do). The survival distribution remains the same at any time period.

The mean residual life expectancy $r(t)$ at age t is given in the table below for men and lightbulbs. It is obtained by subtracting t from the life expectancy at age t.

	Future life expectancy at age t	
Age now t	Male (years)	Lightbulb (time units)
0	73.3	73.3
10	64.2	73.3
20	54.7	73.3
30	45.2	73.3
40	35.9	73.3
50	26.7	73.3
60	18.3	73.3
70	11.5	73.3
80	6.5	73.3
90	3.5	73.3

These results demonstrate that for humans the expected future survival from age t reduces as t increases, but typically for lightbulbs and certain types of electronic components this does not happen. □

There is an interesting alternative expression for (1.9). Begin with

$$r(t) = \frac{\int_t^\infty S(u)du}{S(t)};$$

then, by simplifying the expression for $-\frac{1}{r(t)}$ we find that

$$-\frac{1}{r(t)} = \frac{-S(t)}{\int_t^\infty S(u)du}$$
$$= \frac{\frac{d}{dt}[\int_t^\infty S(u)du]}{\int_t^\infty S(u)du} \qquad (1.12)$$
$$= \frac{d}{dt}[\log_e \int_t^\infty S(u)du].$$

The following theorem indicates that the survival function may be expressed in terms of the mean residual lifetime:

Theorem 1.4 $\quad S(y) = \dfrac{r(0)}{r(y)} e^{-\int_0^y \frac{dt}{r(t)}}$

Proof: Integrate each side of (1.12) between 0 and y to obtain

$$\begin{aligned}
-\int_0^y \frac{dt}{r(t)} &= \int_0^y d[\log_e \int_t^\infty S(u)du] \\
&= \log_e \int_y^\infty S(u)du - \log_e \int_0^\infty S(u)du \\
&= \log_e \int_y^\infty S(u)du - \log_e r(0) \\
&= \log_e \left[\frac{\int_y^\infty S(u)du}{r(0)}\right].
\end{aligned}$$

Exponentiating each side and using (1.9) gives

$$e^{-\int_0^y \frac{dt}{r(t)}} = \frac{\int_y^\infty S(u)du}{r(0)} = \frac{r(y)S(y)}{r(0)}.$$

Finally, the result follows by cross multiplication of each side of the equation by $\frac{r(0)}{r(y)}$. □

1.7 Exercises

1.1. Sketch the survival function for an exponential random variable with mean time to failure of 10 hours. Determine the probability that a component having this survival function will operate at least 20 hours before failure.

1.2. For the aluminium smelter failure data of Example 1.1, estimate the probability of failure beyond 10 weeks.

1.3. Does the function

$$S(u) = \begin{cases} 1-u & \text{for } 0 \leq u \leq 1 \\ 0 & u > 1 \end{cases}$$

satisfy the properties of a survival function for some variable U? What is the distribution of U? Explain briefly.

1.4. Show that a survival random variable Y with hazard function h satisfies

$$P(Y > y + t | Y > y) = \exp[-\int_y^{y+t} h(u)du].$$

If h is an increasing function, is $P(Y > y + t | Y > y)$ an increasing or a decreasing function of y for each $t \geq 0$?

How does this relate to the survival prospects of an 80 year old person?

1.5. Show that if $P(Y > y + t | Y > y)$ is a decreasing function of y for each $t \geq 0$, and the probability density function of Y exists, then $h(y)$ is an increasing function of y.

1.6. Show that if a lifetime variable is exponentially distributed with survival function $S(y) = e^{-\frac{y}{\beta}}$ for $y > 0$ and a positive constant β, then S satisfies

$$S(y+t) = S(y)S(t) \quad \text{for all } y \geq 0, t \geq 0.$$

1.7. It is possible for certain types of components, that once they have successfully functioned for a period of time, they are likely to have a longer life than brand new components (that have not been 'worn in').

If components of a certain type have survival function $S(y)$, then the lifetime distribution is called **new better than used** or **NBU** if
$$S(y+t) \leq S(y)S(t) \quad \text{for all } y \geq 0, t \geq 0.$$
Similarly, the lifetime distribution is termed **new worse than used** or **NWU** if
$$S(y+t) \geq S(y)S(t) \quad \text{for all } y \geq 0, t \geq 0.$$

(a) If lifetimes (in years) are uniformly distributed across the time interval $[0, 3]$, is the distribution NWU or NBU? Explain.

(b) If lifetimes (in years) have a triangular distribution with probability density function
$$f(y) = \begin{cases} \frac{1}{2}y & \text{for } 0 < y < 2 \\ 0 & \text{otherwise,} \end{cases}$$
is the distribution NWU or NBU? Explain.

(c) If lifetimes (in years) have a triangular distribution with probability density function
$$f(y) = \begin{cases} 1 - \frac{1}{2}y & \text{for } 0 < y < 2 \\ 0 & \text{otherwise,} \end{cases}$$
is the distribution NWU or NBU? Explain.

1.8. Determine the hazard function for survival random variable with survival function
$$S(u) = e^{-\frac{1}{2}u^2}.$$

1.9. Let $\alpha > 0$. Determine the survival function for survival random variable with hazard function
$$h(u) = \frac{\alpha u^{\alpha-1}}{1 + u^\alpha}.$$

1.10.(a) Show that if an individual has hazard rate function $h(y)$, then the survival function of the individual is given by
$$S(y) = e^{-\int_0^y h(t)dt}.$$

(b) Show that the probability that an A-year-old individual reaches age B is given by
$$e^{-\int_A^B h(t)dt}.$$

(c) Suppose that the death rate of a person that smokes is, at each age, twice that of a non-smoker.

(i.) If $h_s(y)$ denotes the hazard rate of a smoker at age y and $h_n(y)$ that of a non-smoker at age y, write an equation relating $h_s(y)$ and $h_n(y)$.

(ii.) Use the result of (b) to show that:

Of two individuals of the same age, one of whom is a smoker and the other a non-smoker, the probability that the smoker survives to any given age is the **square** of the corresponding probability of a non-smoker.

(d) The lung cancer hazard rate for a y-year-old male smoker is given by
$$h_s(y) = 0.027 + 0.00025(y - 40)^2, \quad \text{for} \quad y \geq 40.$$
Assuming that a 40-year-old male smoker survives all other hazards, determine the probability that he survives to age 50 without contracting lung cancer?

1.11. If h is the hazard function for a continuous survival random variable, evaluate $\int_0^\infty h(t)dt$.

1.12. Let $Y_1, Y_2, \ldots, Y_n$ be n independent continuous survival random variables with hazard functions $h_1, h_2, \ldots, h_n$. Prove that the survival random variable $Y = \min\{Y_1, Y_2, \ldots, Y_n\}$ has hazard function $h(y) = \sum_{j=1}^n h_j(y)$.

1.13. Let $Y_1, Y_2, \ldots, Y_n$ be n independent continuous survival random variables and define a new survival random variable, Y, by $Y = \min\{Y_1, Y_2, \ldots, Y_n\}$. If Y_i has survival function $S_{Y_i}(y) = \exp(-\alpha_i y^\beta)$ for $\alpha_i > 0$ and $\beta > 0$, determine the hazard function for $n^{\frac{1}{\beta}} Y$.

1.14. Suppose that the time to failure of a cog in an escalator follows an exponential distribution with mean β. The cog will be replaced under a scheduled replacement scheme at fixed time t, even if the cog has not failed by this time. Explain why the time to replacement is given by the variable $Z = \min(Y, t)$. Hence, determine $E(Z)$, the mean time to replacement.

1.15. Let Y be an absolutely continuous survival time with finite mean and survival function S.

(a) Show that $yS(y) \to 0$ as $y \to \infty$.

(b) Hence, use integration by parts to prove that, for fixed t,
$$E(Y|Y > t) = t + \int_t^\infty \frac{S(y)}{S(t)} dy.$$

(c) Evaluate the expected residual lifetime $E(Y - t|Y \geq t)$ when Y has constant hazard.

1.16. If the hazard function of a discrete positive-valued random variable Y is defined by
$$h(y) = \frac{P(Y = y)}{P(Y \geq y)},$$
prove that the Poisson distribution, with rate λ, has a hazard of the form
$$h(y) = \left[1 + \frac{\lambda}{y+1} + \frac{\lambda^2}{(y+1)(y+2)} + \ldots + \frac{\lambda^n}{(y+1)(y+2)\ldots(y+n)} + \ldots\right]^{-1}.$$

Hence, or otherwise, show that
$$1 - \frac{\lambda}{y} < h(y) \leq 1$$
and evaluate $\lim_{y \to \infty} h(y)$.

1.17. If Y is a continuous survival random variable with survival function given by $S(u) = P(Y > u)$, show that
$$E(Y) = \int_0^\infty S(u)du.$$

1.18. If Y is a continuous survival random variable with hazard function $h(y)$ and mean residual lifetime $r(y)$ at time y, show that
$$h(y) = \frac{r'(y) + 1}{r(y)}.$$

1.19. Suppose that Y is a continuous survival random variable with finite mean and survival function $S(y)$. Given that the mean residual lifetime, $r(y)$, at time y is defined by
$$r(y) = \frac{\int_y^\infty S(x)dx}{S(y)},$$
show that
$$-\frac{1}{r(y)} = \frac{d}{dy} \log_e \int_y^\infty S(x)dx;$$
and by integrating each side of this equation, deduce that
$$S(y) = \frac{r(0)}{r(y)} \exp\left(-\int_0^y \frac{du}{r(u)}\right).$$
By differentiating the numerator and denominator of the right-side of $r(y)$, show that
$$\lim_{y \to \infty} r(y) = \lim_{y \to \infty} \left(-\frac{d}{dy} \log_e f(y)\right)^{-1}$$
where $f(y) = -S'(y)$ is the probability density function of Y.
Use this to show that if Y is lognormally distributed, then $r(y) \to \infty$ as $y \to \infty$.

1.20. Show that the mean residual lifetime, $r(t)$, for a survival variable Y with survival function S, may be expressed as
$$r(t) = \int_0^\infty e^{H(t) - H(u+t)} du,$$
in terms of the cumulative hazard function $H(y) = \log_e S(y)$.

EXERCISES

1.21. (*Source*: Zhang and Maguluri, 1994) Consider the following definition based on mean residual lifetimes:

Definition 1.8 *Two survival functions $S_0(y)$ and $S_1(y)$ are said to have* **proportional mean residual lifetime** *(PMRL) if*
$$r_1(y) = \theta r_0(y) \quad \text{for all } y \geq 0, \theta > 0,$$
where $r_0(y)$ and $r_1(y)$ are the respective mean residual lifetimes at time y. □

Show that if S_0 and S_1 have PMRL, then
$$S_1(y) = S_0(y) \left[\int_y^\infty \frac{S_0(u)du}{\mu_0} \right]^{\frac{1}{\theta}-1},$$
where $\mu_0 = r_0(0)$.

Show further that:

(a) if $S_0(y)$ is a survivor function and $r_1(y) = \theta r_0(y)$ for $0 < \theta < 1$ and all $y \geq 0$, then S_1 is a survivor function;

(b) a necessary and sufficient condition for S_1 to be a survivor function for all $\theta > 0$ is that $r_0(y)$ is non-decreasing.

CHAPTER 2

Hazard Models

In Chapter 1 we discovered that survival distributions may be equivalently specified by: f, the probability density function; F, the cumulative distribution function; S, the survival function; h, the hazard function, or r, the mean residual lifetime. This means that if one of these representations is specified, then the others may be derived from it. Therefore, a discussion of lifetime distributions needs only to concentrate on one of these expressions. Often, under such choice, it is the hazard function which is modelled and specified because of its direct interpretation as imminent risk. It may even help identify the mechanism which underlies failures more effectively than the survival function.

Researchers are likely to have first-hand knowledge of the of how imminent risk changes with time for the lifetimes being studied. For example, lightbulbs tend to break quite unexpectedly rather than because they are suffering from old age; people, on the other hand tend to wear out as they get older. We would expect the shapes of the hazard functions for lightbulbs and people to be different. People experience increasing hazard, whereas lightbulbs tend to exhibit constant hazard.

2.1 The constant hazard model

Suppose that $\beta > 0$ is a constant and $\omega = \frac{1}{\beta}$. A **constant hazard model** is usually proposed where imminent risk of failure does not change with time.

$$\textbf{Constant hazard model}: \quad h(y) = \omega \quad \text{for} \quad y > 0. \qquad (2.1)$$

By Theorem 1.3, the corresponding survival function for $y > 0$ is

$$\begin{aligned} S(y) &= e^{-\int_0^y h(u)du} \\ &= e^{-\int_0^y \frac{1}{\beta} du} \\ &= e^{-\frac{1}{\beta} \int_0^y 1 du} \\ &= e^{-\frac{y}{\beta}} \end{aligned}$$

with probability density function $f(y) = -S'(y) = \frac{1}{\beta} e^{-\frac{y}{\beta}}$, mean $E(Y) = \beta$ and variance $Var(Y) = \beta^2$. This characterises the **exponential probability model**. We will use the following notation.

Definition 2.1 *The lifetime variable Y follows an* **exponential probability model** *with mean $\beta > 0$ and we write $Y \sim exp(\beta)$ when the hazard is constant with $h(y) = \frac{1}{\beta}$.* □

This is one of the most commonly used probability models for modelling lifetimes of components; the exponential model often fits survival data well. Why? One possible explanation is that the times between events in a Poisson process are exponentially distributed. Suppose that we are modelling the lifetime of a system. Imagine events (such as toxic shocks to the system) occur according to a Poisson process with rate ω per unit time. Then X_y, the number of shocks occurring in an interval of length y satisfies $X_y \sim Poisson(\omega y)$. If the first such shock is fatal (the system fails), then Y, the time to system failure, satisfies

$$S(y) = P(Y > y) = P(X_y = 0) = \frac{(\omega y)^0 e^{-\omega y}}{0!} = e^{-\omega y},$$

which is the exponential survival function for an exponential random variable with mean $\beta = \frac{1}{\omega}$.

An important feature of the exponential distribution is the 'memoryless property': the future distribution of the lifetime of a component depends on the past only through the present. That is, on reaching any age, the probability of surviving t more units of time is the same as it was at age zero.

Theorem 2.1 *If $Y \sim exp(\beta)$, then for any $y > 0$ and $t > 0$, it follows that $P(Y > y + t | Y > y) = P(Y > t)$.*

Proof: The probability $P(Y > y + t | Y > y)$ may be interpreted as a conditional probability $P(A|B) = P(A \cap B) P(B)^{-1}$, with the identification of the events A and B as $A = \{Y > y + t\}$ and $B = \{Y > y\}$. Then, since $t > 0$, in order that a lifetime be simultaneously longer than both y and $y + t$, it must exceed $y + t$. Therefore, the event $A \cap B$ may be written as $\{Y > y+t\}$. Finally, since $P(Y > r) = e^{-\frac{r}{\beta}}$ for any positive value of r,

$$\begin{aligned} P(Y > y+t | Y > y) &= \frac{P(\{Y > y+t\} \cap \{Y > y\})}{P(\{Y > y\})} \\ &= \frac{P(\{Y > y+t\})}{P(\{Y > y\})} \\ &= \frac{e^{-\frac{y+t}{\beta}}}{e^{-\frac{y}{\beta}}} \\ &= e^{-\frac{t}{\beta}} \\ &= S(t). \end{aligned}$$

□

Theorem 2.1 states that for a component with an exponentially distributed lifetime, the probability that a 3-month-old component lasts 3 more weeks

THE CONSTANT HAZARD MODEL

in operation is the same as the probability that a 6-month-old component lasts 3 more weeks in operation. This indicates that the component's lifetime does not pass through a period of 'old age', where there is an increased risk of mortality. In this sense, an exponential lifetime is different from a human lifetime, where the survival distribution shrinks with age.

■ Example 2.1
What does constant hazard data look like?

Simulated exponential data from a distribution with mean 100 appears in the stem-and-leaf plot below. Two hundred data points are represented: the size of each leaf unit is 10; the stems have increments of 20.

```
              Stem-and-leaf plot                   Estimated hazard

    31     0 000000000000000011111111111111           0.0078
    61     0 222222222222222222223333333333            0.0089
    84     0 4444444444444444455555555                 0.0083
   (27)    0 6666666666667777777777777                 0.0116
    89     0 888888899                                 0.0051
    80     1 000000001111111                           0.0094
    65     1 22223333333                               0.0085
    54     1 4445555                                   0.0065
    47     1 666666777                                 0.0096
    38     1 88899                                     0.0067
    33     2 00111111                                  0.0121
    25     2 33333                                     0.0100
    20     2 455                                       0.0075
    17     2 677                                       0.0088
    14     2 8                                           ↓
    13     3 11
    11     3 22                                    w = 0.01 = 1/100
     9     3 5
     8     3 6
     7     3 9
     6     4 0
     5     4 2
     4     4 45
     2     4
     2     4
     2     5
     2     5 2
     1     5 4
```

These data have an actual mean of 107.47 and a standard deviation of 106.13, reflecting that the mean and standard deviation are the same for the exponential lifetime model. In Chapter 4 we will examine tests and plots to determine whether a given data set fits the exponential model for lifetimes.

However at this stage, it is important to note the general shape of the plot, its extreme right-skewness and the presence of high outlying values exceeding five times the mean.

The right-hand column beside the stem-and-leaf plot shows simple estimates of the hazard function based on counting the data in the stem-and-leaf plot. Using an intuitive hazard estimate (see Chapter 1) based on the formula

$$\frac{P(y < Y < y + \Delta y)}{(\Delta y) P(Y > y)}$$

for average failure rate across intervals of size $\Delta y = 20$, we obtain, for example, in the first line, a hazard estimate of

$$\frac{\frac{31}{200}}{20 \frac{200}{200}} = 0.00775.$$

There are 31 observations in the first line of the plot and 200 observations lie in or beyond the first line of the plot. Therefore, for $y = 0$, we have taken $\frac{31}{200}$ as an estimate of $P(y < Y < y + \Delta y)$ and $\frac{200}{200}$ as an estimate of $P(Y > y)$. The calculations are similar for the other lines of the plot. For example, in the second line of the plot, the hazard estimate is

$$\frac{\frac{30}{200}}{20 \frac{169}{200}} = 0.00888.$$

It must be emphasised that this is an elementary rule-of-thumb way of estimating the hazard function from a stem-and-leaf plot. Notice that nearly all the estimates of the hazard function in the right-hand column are 0.01 to two decimal places, suggesting the hazard function $h(y) = \frac{1}{\beta}$ for $\beta = 100$. □

The parameter β is very importantly a scale parameter. It is easy to establish that the following theorem holds:

Theorem 2.2 *If $Y \sim exp(\beta)$, then $\frac{Y}{\beta} \sim exp(1)$.* □

In particular, this theorem suggests that if we multiply or divide exponential lifetimes by a constant, then the MTTF correspondingly multiplies or divides by the same constant.

2.2 The power hazard model

Suppose that $\alpha > 0$ and $\beta > 0$ are constants. A **power hazard model** is usually proposed where imminent risk of failure is rapidly increasing with time.

Power hazard model: $\quad h(y) = \frac{\alpha}{\beta^\alpha} y^{\alpha-1} \quad$ for $y > 0$. $\quad$ (2.2)

Notice that the form of (2.2) is simply $h(y) = (constant) y^{constant}$; however the choice of the parametrisation of the power hazard in terms of α and β

becomes clear when the survival function is determined. By Theorem 1.3, the corresponding survival function is, for $y > 0$,

$$\begin{aligned} S(y) &= e^{-\int_0^y h(u)du} \\ &= e^{-\int_0^y \frac{\alpha}{\beta^\alpha} u^{\alpha-1} du} \\ &= e^{-\frac{1}{\beta^\alpha} \int_0^y \alpha u^{\alpha-1} du} \\ &= e^{-(\frac{y}{\beta})^\alpha}. \end{aligned}$$

Therefore the survival function for a variable with power hazard has a particularly simple form where the power is translated to the exponent:

$$S(y) = e^{-(\frac{y}{\beta})^\alpha}.$$

The probability density function (pdf) $f(y) = -S'(y)$ is

$$f(y) = \frac{\alpha}{\beta^\alpha} y^{\alpha-1} e^{-(\frac{y}{\beta})^\alpha}, \quad y > 0.$$

Definition 2.2 *The lifetime random variable Y follows a* **Weibull probability model** *with parameters $\alpha > 0$ and $\beta > 0$ and we write $Y \sim Weibull(\alpha, \beta)$ when Y has a power hazard of the form $h(y) = \frac{\alpha}{\beta^\alpha} y^{\alpha-1}$.* □

As in the exponential case, the parameter β is very importantly a scale parameter. The exponent of the survival function has the response y divided by β. It is easy to establish that the following theorem holds.

Theorem 2.3 *If $Y \sim Weibull(\alpha, \beta)$, then $\frac{Y}{\beta} \sim Weibull(\alpha, 1)$.* □

The Weibull model has a very simple hazard function and a simple closed form survival function. These, along with its two parameter flexibility, make it a very useful model in many engineering contexts. However, the mean and variance of the distribution are somewhat more difficult to determine. For this we will need the 'gamma function'.

Definition 2.3 *The* **gamma function**, $\Gamma(\alpha)$, *is defined for all $\alpha > 0$ by the integral $\Gamma(\alpha) = \int_0^\infty t^{\alpha-1} e^{-t} dt$.* □

The gamma function enjoys many interesting properties and special features which we outline in the following theorem. The establishment of 1 and 2 are discussed in the problems at the end of this section; 3 follows through integration by parts; and 4 is the result of repeated application of 3 using $x = n$.

Theorem 2.4 *Properties of the gamma function.*
1. $\Gamma(1) = 1$.
2. $\Gamma(\frac{1}{2}) = \sqrt{\pi}$.
3. $\Gamma(x+1) = x\Gamma(x)$, *for x any positive real number.*
4. $\Gamma(n+1) = n!$, *for $n = 1, 2, 3, \ldots$, a positive integer.* □

The gamma function is used to determine the Weibull mean and variance. First, we take advantage of a scale change: using the distribution of $X \sim Weibull(\alpha, 1)$, the moments about 0 are given by

$$E(X^r) = \int_0^\infty x^r \alpha x^{\alpha-1} e^{-x^\alpha} dx$$
$$= \int_0^\infty u^{\frac{r}{\alpha}} e^{-u} du \quad (\text{substitution } u = x^\alpha)$$
$$= \Gamma(1 + \frac{r}{\alpha}).$$

Now if $Y = \beta X \sim Weibull(\alpha, \beta)$, then the rth moment about 0 is given by

$$E(Y^r) = E(\beta^r X^r) = \beta^r E(X^r) \doteq \beta^r \Gamma(1 + \frac{r}{\alpha}).$$

The first and second moments about 0 are used to write the mean $E(Y)$ and variance $E(Y^2) - E(Y)^2$ of Y.

Theorem 2.5 *If $Y \sim Weibull(\alpha, \beta)$, then $E(Y) = \beta \Gamma(1 + \frac{1}{\alpha})$ and $Var(Y) = \beta^2 [\Gamma(1 + \frac{2}{\alpha}) - \Gamma(1 + \frac{1}{\alpha})^2]$.* □

What models may be constructed for different values of the parameter α? When $\alpha = 1$, the Weibull model reduces to an exponential model with constant hazard; that is, $Weibull(1, \beta) \equiv exp(\beta)$. If $\alpha > 1$, then the Weibull hazard function is increasing. Similarly, if $\alpha < 1$, then the Weibull hazard function is decreasing. This makes the Weibull a very flexible model in a wide variety of situations: increasing hazards, decreasing hazards, constant hazards.

■ **Example 2.2**
Weibull batteries
Suppose that the length of life, Y, in years, for a car battery has a probability density function

$$f_Y(y) = \begin{cases} \frac{2}{3} y e^{-\frac{y^2}{3}}, & \text{if } y > 0 \\ 0, & \text{otherwise}. \end{cases}$$

We demonstrate how to identify this member of the Weibull family and determine the hazard function, reliability function, mean and standard deviation, and probability that a battery will last at least 4 years given that it is now 2 years old.

For these questions, a direct exponent by exponent comparison of the stipulated probability density function (pdf) with the general Weibull pdf

$$f_X(x) = \begin{cases} \frac{\alpha}{\beta^\alpha} x^{\alpha-1} e^{-(\frac{x}{\beta})^\alpha}, & \text{if } x > 0; \\ 0, & \text{otherwise} \end{cases}$$

for $X \sim Weibull(\alpha, \beta)$, shows that $Y \sim Weibull(2, \sqrt{3})$. That is, $\alpha = 2$ and

THE POWER HAZARD MODEL

$\beta = \sqrt{3}$. Therefore,

$$\mu = E(Y) = \sqrt{3}\Gamma(1 + \tfrac{1}{2}) = \sqrt{3}[\tfrac{1}{2}\Gamma(\tfrac{1}{2})] = 1.53 \quad (2 \text{ dp})$$

and

$$\sigma = SD(Y) = \sqrt{3}[\Gamma(2) - \Gamma(1 + \tfrac{1}{2})^2]^{\tfrac{1}{2}} = \sqrt{3}[1 - \tfrac{\pi}{4}]^{\tfrac{1}{2}} = 0.80 \quad (2 \text{ dp}),$$

and the survival function is simply $S_Y(y) = e^{-\tfrac{y^2}{3}}$, $y > 0$. The required probability is then $P(Y \geq 4 | Y \geq 2)$ which may be evaluated as the conditional probability of the event $\{Y \geq 4\}$ given the event $\{Y \geq 2\}$. Using the expression just derived for the reliability function,

$$P(Y \geq 4 | Y \geq 2) = \frac{P(Y \geq 4)}{P(Y \geq 2)} = \frac{e^{-\tfrac{4^2}{3}}}{e^{-\tfrac{2^2}{3}}} = e^{-4} = 0.018 \quad (3 \text{ dp}).$$

The size of this probability is not unexpected, given the size of the mean and variance of Y. Finally note that the hazard function is linear:

$$h_Y(y) = \frac{2}{3}y, \quad \text{for} \quad y > 0.$$

Under this Weibull model, the batteries show a linearly increasing risk of failure.

In the stem-and-leaf plot below we use statistical software to simulate 200 observations from the $Weibull(2, \sqrt{3})$ model. The leaf units are of size 0.1 and the increment is 0.5 between adjacent stems.

```
  16    0  1122333333344444
  56    0  55555566666666667777777778888888888999999
 (62)   1  000000000000011111111111111111222222222233333344444444444444
  82    1  5555555566666666666677777788888888888999
  43    2  000000000111223334444
  21    2  55566666788999
   7    3  124
   4    3  789
   1    4  4
```

Notice the shape of this stem-and-leaf plot; the right-skewness shows in a longer tail toward higher values. Following the method of estimation described in Example 2.1, the hazard estimates based on

$$\frac{P(y < Y < y + \Delta y)}{(\Delta y) P(Y > y)}$$

for the first few lines of the stem-and-leaf plot are (using $\Delta = 0.5$), 0.16, 0.43, 0.86, 0.95, 1.02, 1.3. Clearly, the hazard is increasing with a hint of linear trend. □

When power hazards are added, polynomial hazards are constructed; suppose that $a_0, a_1, \ldots, a_k$ are constants.

Rayleigh hazard model : $h(y) = a_0 + a_1 y + a_2 y^2 + \ldots + a_k y^k.$ (2.3)

Such hazards have been usefully modelled for $k = 1$. For $k = 2$, the coefficients could be chosen so that $h(y)$ follows a quadratic curve or a 'bathtub' shape (with $h(y) > 0$ for all $y > 0$) as outlined in Section 2.5. See the exercises in Chapter 1 for calculations based on the 'smoker's hazard' $h_s(y) = 0.027 + 0.00025(y - 40)^2$ for $y \geq 40$.

2.3 Exponential hazards

We have discussed simple forms for the hazard function, including constant, power and linear. Exponential hazards provide for a natural extension, especially since the survival function still remains easy to calculate with a closed-form formula. Suppose that $b > 0$ and $-\infty < u < \infty$ are constants.

Exponential hazard model : $h(y) = \frac{1}{b} \exp(\frac{y-u}{b}), 0 < y < \infty.$ (2.4)

Notice that the form of (2.4) is simply $h(x) = (constant)e^{constant}$. An **exponential hazard model** occurs frequently in actuarial science for modelling human lifetimes. By Theorem 1.3, the survival function is, for $y > 0$,

$$\begin{aligned} S(y) &= e^{-\int_0^y h(t)dt} \\ &= \exp\left[-\int_0^y \frac{1}{b} e^{\frac{t-u}{b}} dt\right] \\ &= \exp\left[-e^{\frac{t-u}{b}} \big|_0^y\right] \\ &= \exp\left[-\exp(\frac{y-u}{b}) + \exp(-\frac{u}{b})\right]. \end{aligned}$$

If this survival function is carefully re-labelled (or re-parametrised), it may be expressed in the form

$$S(y) = \exp\left[\theta(1 - e^{\alpha y})\right],$$

advocated as the **Gompertz survival model** by Moeschberger and Klein (1997). The parametric relationships between (α, θ) and (u, b) are

$$\alpha = \frac{1}{b} \quad \text{and} \quad \theta = e^{-\frac{u}{b}}.$$

Notice that the probability density function is given by

$$f(y) = -S'(y) = \frac{1}{b} \exp[\frac{y-u}{b} - \exp(\frac{y-u}{b})] = \frac{1}{b} \exp(\frac{y-u}{b}) S(y).$$

Definition 2.4 *The lifetime random variable Y follows a **Gompertz probability model** with parameters $b > 0$ and $-\infty < u < \infty$, and we write $Y \sim Gompertz(u, b)$ when Y has an exponential hazard of the form $h(y) = \frac{1}{b} \exp(\frac{y-u}{b}), 0 < y < \infty$.* □

EXPONENTIAL HAZARDS

A 'lifetime model' closely related to the Gompertz is the so-called **extreme value probability model**. Its close relationship with the Gompertz model is very close indeed: they have the same formula for the hazard function, however the extreme value hazard is defined across the whole real line $-\infty < y < \infty$ and not just $y > 0$. Because negative values can occur in the extreme value distribution, we should not officially call it a survival distribution. Its importance, however, is that it arises as the logarithm of a survival distribution, and frequently survival data may be quite naturally described on the log scale.

The expansion of the domain of definition of the Gompertz hazard to the whole real line means that the parameters may now be described through a **location parameter** u and **scale parameter** b. This becomes clear as we develop the extreme value model through transforming the Weibull distribution: suppose that $T \sim Weibull(\alpha, \beta)$. We will derive the survival function $S(y)$ of $Y = \log_e T$ for $-\infty < y < \infty$:

$$\begin{aligned}
S(y) &= P(Y > y) \\
&= P(\log_e T > y) \\
&= P(T > e^y) \\
&= \exp[-(\frac{e^y}{\beta})^\alpha] \\
&= \exp[-(\frac{1}{\beta})^\alpha e^{\alpha y}] \\
&= \exp[-\exp(\alpha y - \alpha \log_e \beta)] \\
&= \exp[-\exp(\frac{y-u}{b})],
\end{aligned}$$

where $b = \frac{1}{\alpha}$ and $u = \log_e \beta$. This choice of parametrization means that $b > 0$ and $-\infty < u < \infty$. Notice that the probability density function for Y is given by

$$f(y) = -S'(y) = \frac{1}{b}\exp[\frac{y-u}{b} - \exp(\frac{y-u}{b})] = \frac{1}{b}\exp(\frac{y-u}{b})S(y)$$

for $-\infty < y < \infty$.

Definition 2.5 *The random variable Y follows an* **extreme value probability model** *and we write $Y \sim extremevalue(u, b)$ when $h(y)$ is of the form $h(y) = \frac{1}{b}\exp(\frac{y-u}{b})$ for $-\infty < y < \infty$ and parameters $b > 0$ and $-\infty < u < \infty$.*
□

The effect of the increased domain for the extreme value distribution may be seen by comparing the extreme value survival function

$$S_1(y) = \exp[-\exp(\frac{y-u}{b})], \quad -\infty < y < \infty$$

with the Gompertz survival function

$$S_2(y) = \exp[-\exp(\frac{y-u}{b}) + \exp(-\frac{u}{b})], \quad y > 0.$$

The extra exponent term in S_2 is the normalising factor to ensure that the area under the corresponding density is 1. The location-shift and scale-change of the extreme value distribution may be expressed through the linear relationship

$$Y = u + bX,$$

where $Y \sim extremevalue(u,b)$ and $X \sim extremevalue(0,1)$. Clearly, this linear transformation property does not hold for the Gompertz distribution.

The **standard extreme value probability model** has parameter specifications $u = 0$ and $b = 1$. We examine properties of the standard extreme value model since properties in the non-standard case are obtained by linear transformation as above. To this end, suppose that $Y \sim extremevalue(0,1)$. Then Y has probability density function $f(y) = \exp[y - \exp(y)]$, and moment generating function (to yield the first few moments of the distribution — mean and variance):

$$\begin{aligned} m(t) &= E(e^{tY}) \\ &= \int_{-\infty}^{\infty} e^{ty} f(y) dy \\ &= \int_{-\infty}^{\infty} e^{ty} \exp[y - e^y] dy \\ &= \int_0^{\infty} u^t e^{-u} du \quad \text{for} \quad u = e^y \\ &= \Gamma(1+t). \end{aligned}$$

From this suprisingly simple moment generating function, the mean and variance are readily extracted in terms of the not-so-well-known Euler's constant $\gamma = 0.5772$ (4 dp). The mean and variance are found by

$$E(Y) = m_Y'(0) = \frac{d}{dt}\Gamma(1+t)|_{t=0} = \Gamma'(1) = -\gamma$$

and

$$Var(Y) = \Gamma''(1) - \gamma^2 = \frac{\pi^2}{6}.$$

2.4 Other hazards

Since many lifetimes are measured on the logarithm scale, and such transformations often increase the symmetry in data, it is important to examine the **lognormal distribution** where the transformed data are normally distributed.

OTHER HAZARDS

Definition 2.6 *A positive valued survival variable T has a **lognormal distribution** and we write $T \sim lognormal(\mu, \sigma)$ if $X = \log_e T \sim N(\mu, \sigma^2)$. That is, X is normally distributed with mean μ and variance σ^2.* □

The survival function and the hazard function for the lognormal distribution do not have simple representations. However, if Φ is the cumulative distribution function of the standard normal random variable, then $T \sim lognormal(\mu, \sigma)$ has survival function

$$S(t) = 1 - P(T < t) = 1 - P(\log_e T < \log_e t) = 1 - \Phi(\frac{\log_e t - \mu}{\sigma}),$$

from which the density and hazard may be obtained by differentiation. The major difficulty with the lognormal distribution is that the hazard rate is like a skewed mound in shape: it initially increases, reaches a maximum and then decreases toward 0 as lifetimes become larger and larger. This is hardly the stuff of lifetime modelling where hazards increase with old age! However, authors such as Horner (1987) have used the lognormal to model the onset of Alzheimer's disease utilising the left tail of the distribution.

There are other hazard models that mimic normal distribution properties with closed-form hazard and survival functions. One example is the **log-logistic distribution**.

Definition 2.7 *A positive valued survival variable Y has a **log-logistic distribution** and we write $Y \sim loglogistic(\mu, \sigma)$ with $\sigma > 0$ and $-\infty < \mu < \infty$ if $X = \log_e Y$ follows a logistic distribution with survival function*

$$S_X(x) = 1 - \frac{1}{1 + \exp[-(\frac{x-\mu}{\sigma})]}.$$

□

The survival function of the log-logistic distribution is therefore

$$\begin{aligned} S_Y(y) &= P(Y > y) \\ &= P(X > \log_e y) \\ &= S_X(\log_e y) \\ &= 1 - \frac{1}{1 + \exp[-(\frac{\log_e y - \mu}{\sigma})]} \\ &= \frac{1}{1 + \theta y^\alpha}, \end{aligned}$$

where $\alpha = \frac{1}{\sigma} > 0$ and $\theta = \exp(-\frac{\mu}{\sigma})$. The hazard function

$$h_Y(y) = \frac{\alpha \theta y^{\alpha-1}}{1 + \theta y^\alpha}$$

derived from S_Y is similar in shape to the lognormal hazard, but is considerably easier to manipulate.

2.5 IFR

The process of ageing is measured by the hazard function. For example, humans have a greatly increased imminent risk of death as they reach old age and this corresponds to an increasing hazard function.

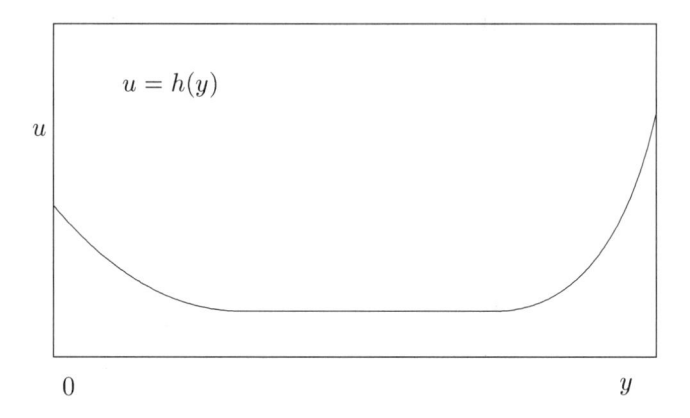

Figure 2.1 *Diagram showing the graph of a bathtub hazard characteristic of human lifetimes: for newborns, the imminent risk if death decreases with time; in middle age, the hazard is relatively constant; finally, the hazard rate increases during old age.*

Definition 2.8 *The random variable Y has* **increasing failure rate** *or* **IFR** *(respectively* **decreasing failure rate** *or* **DFR***) if $h(y)$, the hazard function of Y, is an increasing (respectively decreasing) function of y.* □

Since constant failure rate corresponds to the exponential model, an IFR forebodes worsening survival prospects through the equation

$$P(Y > y + t | Y > y) = \exp[-\int_y^{y+t} h(u)du],$$

established in the exercises of Chapter 1, which shows directly the conditional survival probability for a y-year-old decreasing as a function of y for each $t > 0$ when $h(y)$ is increasing.

Humans in infancy experience DFR from the initial hurdle of birth. Infant mortality decreases with time during the immature period. Similarly, manufactured items may experience a wear-in period. Adulthood is typified by a reasonably constant hazard, which increases beyond the age of 40 (the wear-out phase). This creates overall a **bathtub-shaped hazard** across the support of the lifetime distribution as illustrated in Figure 2.1. Note that if $Y \sim Weibull(\alpha, \beta)$, then Y has IFR if $\alpha > 1$, DFR if $\alpha < 1$ and constant hazard if $\alpha = 1$.

EXERCISES 31

The final example of this section concerns a new distribution which is clearly DFR as the hazard decreases for values beyond a fixed constant. This is hardly a model for human lifetimes!

Definition 2.9 *A positive-valued survival variable has a* **Pareto distribution** *and we write* $Y \sim Pareto(\alpha, \gamma)$ *with* $\alpha > 0$ *and* $\gamma > 0$ *if* Y *has hazard function*

$$h(y) = \frac{\alpha}{y} \quad \text{for } y \geq \gamma.$$

□

The support of this distribution is $[\gamma, \infty)$. The parameter γ acts as a threshold value beyond which data may be measured. This applies particularly to certain types of astronomical data where distant objects cannot be detected if they are not of sufficient magnitude.

2.6 Exercises

2.1. Identify the distribution which has hazard function:

(a) $h(y) = 1$;
(b) $h(y) = y$;
(c) $h(y) = y^2$;
(d) $h(y) = (1+y)^2$;
(e) $h(y) = e^y$ for $y > 0$;
(f) $h(y) = e^y$ for $-\infty < y < \infty$;
(g) $h(y) = \frac{1}{1+y}$.

State the survival function in each case. Which of these distributions is IFR?

2.2. Write down the survival function, hazard function and probability density function for $Y \sim Weibull(\frac{1}{2}, 4)$.

2.3. Show that if $Y \sim Weibull(\alpha, \beta)$, then Y has IFR if $\alpha > 1$, DFR if $\alpha < 1$ and constant hazard if $\alpha = 1$.

2.4. Past age 40, the hazard curve for an individual's survival is nearly linear when plotted on a logarithmic scale. This suggests that the hazard is exponential:

$$h(y) = ae^{b(y-40)} \quad \text{for } y > 40.$$

Determine the survival function over this range.

What shape would you expect for the hazard function for people aged under 40?

2.5. Nelson (*Journal of Quality Technology*, July 1985) suggests that the Weibull distribution usually provides a better representation for the lifelength of a product than the exponential distribution. Nelson used a Weibull distribution with $\alpha = 1.5$ and $\beta = 23$ to model the lifelength of a roller bearing (in thousands of hours).

(a) Find the probability that a roller bearing of this type will have a service life of less than 12.2 thousand hours.

(b) Show that a Weibull distribution with $\alpha = 1$ is an exponential distribution. Nelson claims that very few products have an exponential life distribution, although such a distribution is commonly applied. Calculate the probability from part (a) using the exponential distibution. Compare your answer with that obtained in part (a).

2.6. Let X denote the corrosion weight loss for a small square magnesium alloy plate immersed for 7 days in an inhibited aqueous 20% solution of $MgBr_2$. Suppose that the minimum possible weight loss is 3 units and that the excess $Y = X - 3$ over this minimum has a Weibull distribution with density function
$$f_Y(y) = \frac{1}{8} y e^{-\frac{y^2}{16}}$$
for $y \geq 0$. Determine $P(7 \leq X \leq 9)$.

2.7. According to Definition 2.9, the Pareto distribution with $\alpha > 0$ and $\gamma > 0$ has hazard function
$$h(y) = \frac{\alpha}{y} \text{ for } y \geq \gamma.$$

(a) Sketch the shape of this hazard function.

(b) Determine the Pareto survival function and density function in terms of the parameters α and γ.

2.8. The survival times (in days after transplant) for the original $n = 69$ members of the Stanford Heart Transplant Program (see Crowley and Hu, 1977) were as follows:

Survival time after heart transplant (days)							
15	3	624	46	127	64	1350	280
23	10	1024	39	730	136	1775	1
836	60	1536	1549	54	47	51	1367
1264	44	994	51	1106	897	253	147
51	875	322	838	65	815	551	66
228	65	660	25	589	592	63	12
499	305	29	456	439	48	297	389
50	339	68	26	30	237	161	14
167	110	13	1	1			

(Source: Crowley, J. and Hu, M. (1977). The covariance analysis of heart transplant data. *Journal of the American Statistical Association*, 72, 27–36.)

Construct a stem-and-leaf plot for these data on the $\log_{10}$ scale. Use the plot to provide simple estimates of the hazard function against each stem in the plot.

EXERCISES

2.9. Show that if $Y \sim Weibull(\alpha, \beta)$, then
$$U = \left(\frac{Y}{\beta}\right)^\alpha \sim exp(1).$$

2.10. Recall that the random variable Y follows a log logistic model if $Y \sim loglogistic(\mu, \sigma)$. For what range of values of $\alpha = \frac{1}{\sigma}$ is this model IFR? DFR? Explain.

2.11. (**Expected service life**). Suppose that Y denotes the 'time to failure' of a component in manufacturing equipment. A replacement policy designed to avoid excess cost in replacing failed components ensures that components are replaced regularly (hopefully before they fail). The expected service life, $ESL(t)$, of the component when it is automatically replaced after t hours of usage is defined as the expected value of the mixture random variable
$$Z = \min\{Y, t\}.$$

(a) Explain heuristically why Z represents the 'service life' of a component.

(b) Z may be regarded as a function of the random variable Y with a single point discontinuity at t. Let f denote the pdf of Y. Use the methods for determining the expected value of a function of a random variable, namely
$$E[g(Y)] = \int_0^\infty g(y) f(y) dy,$$
to show that
$$ESL(t) = \int_0^t y f(y) dy + t P(Y > t).$$
[This is the formula implied by Brick, Michael and Morganstein (1989) in their paper on using statistical thinking to solve maintenance problems.]

(c) Determine $ESL(2)$ for a 2-year service life policy for batteries where $Y \sim Weibull(2, \sqrt{3})$.

2.12. Assume that for a certain grade of steel beam, the yield strength Y, in kg mm^{-2}, follows a Weibull distribution with parameters $\alpha = 3$ and $\beta = 8$ for values above a threshold level of $t_0 = 30$ kg mm^{-2}. This means that the Weibull distribution applies to values $W = Y - 30$, so we assume that $W = Y - 30 \sim Weibull(3, 8)$.

(a) Determine $P(8 < W < 10)$.

(b) Determine the probability that the yield strength is between 38 and 46 kg mm^{-2}.

(c) Determine the mean and variance in yield strength.

(d) Suppose that the threshold value t_0 for a particular grade of steel is unknown. How would you propose to estimate the threshold from 100 observations of yield strength? For example, would the average, maximum or minimum of the observations provide an appropriate estimate of the threshold?

(e) The threshold value t_0 is often built into the Weibull distribution as an extra parameter to create the three parameter Weibull distribution. From
$$W = Y - t_0 \sim Weibull(\alpha, \beta),$$
determine the cumulative distribution function (cdf) of Y, and by differentiating this expression, determine the pdf of Y.

2.13. If Y has a Weibull distribution with standard parameters α and β, determine the probability density function of $X = \log Y$. Determine the moment generating function of X in the case that $\alpha = 1$ and $\beta = 1$. Use this moment generating function to evaluate $E(X)$.

2.14. In this problem we examine the location-shift and scale-change of the extreme value distribution. Show that if $X \sim extremevalue(0, 1)$ and
$$Y = \mu + \sigma X,$$
for constants $\sigma > 0$ and $-\infty < \mu < \infty$, then $Y \sim extremevalue(\mu, \sigma)$.

2.15. Suppose that a population contains individuals for which lifetimes Y are exponentially distributed, but that the rate varies across individuals. Specifically, suppose that the distribution of Y given λ has probability density function $f(y \mid \lambda) = \lambda e^{-\lambda y}$ for $y \geq 0$, and that λ itself has a gamma distribution with probability density function
$$g(\lambda) = \frac{\lambda^{k-1} e^{-(\lambda/\alpha)}}{\alpha^k \Gamma(k)},$$
for $k > 0$.

(a) Find the unconditional probability density function and survival function for Y.

(b) Show that the unconditional hazard function is
$$h(y) = \frac{k\alpha}{1 + \alpha y}.$$

(c) Show that h is monotone decreasing.

(d) Show that individuals (components) having survival function Y in laboratory tests, when put into service where they are subjected to possible catastrophic accidents which have a constant rate of occurrence θ independent of the failure time distribution represented by Y, will have a hazard function of the form
$$h(y) = \frac{k\alpha}{1 + \alpha y} + k\alpha\theta.$$
This can be regarded as a failure model in which both infant mortality and the constant mortality rate of mid-life accidents are present but there is no evidence of wear-out (i.e., there are no terminally increasing hazard rates). This occurs for electronic equipment containing

2.16. Consider the problem of designing a life test in which 20 electric motors are run to destruction, giving independent observations $Y_1, Y_2, \ldots, Y_{20}$ (measured in hours) from a $Weibull(\frac{1}{2}, 10)$ distribution. (This distribution may be suggested by previous test results for motors of a similar design.) Because of time and cost considerations, the test cannot run indefinitely.

(a) Estimate the test fixed duration, t, which should ensure that 95% of the motors will have burned out by the test termination date? [*Hint*: Determine the upper 5% point of the Weibull distribution.]

(b) Determine the probability that a particular motor, perhaps the seventeenth one, fails before t. That is, find $P(Y_{17} < t)$.

(c) Determine $P(Y_i < t)$ for each $i = 1, 2, 3, \ldots, 20$.

(d) Determine the probability that all the 20 test motors fail before t.

(e) Give an expression for the probability that the first motor to fail (from amongst the 20 on test) does so after t.

(f) An observation of the failure time of a test motor is termed *censored* if the motor does not fail before the test termination date t. Determine the probability that 2 of the 20 motors are censored by the fixed test termination date.

2.17. (**Characterising the exponential distribution.**) Suppose that a survival random variable Y has survival function S which satisfies

$$S(y+t) = S(y)S(t) \quad \text{for all} \quad y > 0, \ t > 0.$$

This question shows that there exists a constant ω such that $Y \sim exp(\frac{1}{\omega})$.

(a) Let $c > 0$ and m and n be positive integers. Show that

$$S(nc) = [S(c)]^n \quad \text{and} \quad S(c) = [S(\frac{c}{m})]^m.$$

(b) Use (a) to show that $0 < S(1) < 1$. [Try assuming that $S(1) = 0$ and $S(1) = 1$ and produce contradictions in each case.]

(c) By (b), there exists $0 < \omega < \infty$ such that $S(1) = e^{-\omega}$. Use this to establish that:

(i.) $S(\frac{1}{m}) = e^{-\frac{\omega}{m}}$,
(ii.) $S(\frac{n}{m}) = e^{-\frac{n\omega}{m}}$ and
(iii.) $S(y) = e^{-\omega y}$ for all $y > 0$.

2.18. If $Y \sim exp(1)$, show that the moment generating function, $m_Y(t)$, of Y is given by

$$m_Y(t) = E(e^{tY}) = \frac{1}{1-t}, \quad \text{for} \quad t < 1.$$

By expanding this expression as a power series, write the exponential moments $E(Y^k)$ for $k = 0, 1, 2, 3, \ldots$

2.19. For $k > 0$, the gamma function $\Gamma(k)$ is defined by $\Gamma(k) = \int_0^\infty u^{k-1} e^{-u}\, du$. Show that:

(a) $\Gamma(1) = 1$,
(b) $\Gamma(\frac{1}{2}) = \sqrt{\pi}$ and
(c) $\Gamma(k+1) = k\Gamma(k)$.

2.20. (**Different parameters for the Weibull distribution.**) A Weibull random variable Y with **rate** λ and **Weibull index** γ has probability density function
$$f(y) = \lambda\gamma(\lambda y)^{\gamma-1} e^{-(\lambda y)^\gamma}.$$

This is simply a reparametrisation of the Weibull model of (2.2). In terms of λ and γ, determine $E(Y), E(Y^2)$, and, in general, $E(Y^r)$ for r a positive integer.

Deduce a simple expression for $E(Y^r)$ in the case that γ is a positive integer and $r = k\gamma$ for some positive integer k.

CHAPTER 3

Reliability of Systems

We have considered modelling the lifetimes of components, people, units... by using hazards and survival functions. However, it is especially true for engineering applications that pieces of equipment consist of many (possibly different) interacting components. The term **reliability** is commonly used to describe the 'survival' of such components and of such a system. Essentially, the reliability of a component is the probability that it is operational. In this chapter we consider how the reliability of individual components affects the reliability of the entire system. How does the individual affect the whole? A major reference on reliability is Barlow and Proschan (1975). Their discussion is more extensive than that presented here.

3.1 Coherent systems and structure functions

It is certainly true that the reliability of components may change with time, and in the next section we will examine time dependence on reliability. However, initially let us make the assumption that at some instant in time we are able to observe a component and know whether it is functioning or not functioning. Since we will be considering n components and their inter-relationships, it will be helpful to label and list them:

$$C1, C2, C3, \ldots, Cn.$$

In general, Ci represents component i. Now suppose that each component has one of two operational states; 'functioning' and 'not functioning'. We can record the state of a component using an indicator function: for each $i = 1, 2, 3, \ldots, n$, the indicator X_i associated with Ci is defined by

$$X_i = \begin{cases} 1 & \text{if } Ci \text{ is functioning} \\ 0 & \text{if } Ci \text{ is not functioning.} \end{cases}$$

The indicator function for a system of components is termed a 'structure function'.

Definition 3.1 *The* **structure function** *for a system of n components with indicators $X_1, X_2, X_3, \ldots, X_n$ is given by*

$$\phi(X_1, X_2, X_3, \ldots, X_n) = \begin{cases} 1 & \text{if the system is functioning} \\ 0 & \text{if the system is not functioning.} \end{cases}$$

The number n of components is termed the **order of the system**. □

We can see how structure functions describe systems by looking at a few simple examples.

By a **series system** we mean a configuration of n components as illustrated in Figure 3.1.

Figure 3.1 *A configuration of components $C1, C2, \ldots, Cn$ linked in series. For successful operation, the current flows from left to right through the components. Clearly, when a component fails, the system is not able to function.*

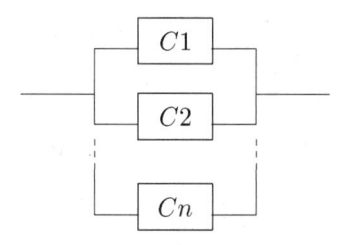

Figure 3.2 *A configuration of components $C1, C2, \ldots, Cn$ linked in parallel. For successful operation, the current flows from left to right through the components. Clearly, when a component fails, the system is still able to function.*

A series system is not functioning if one (or more) of the components is not functioning. It is easy to verify that for a series system, because each of the indicator functions X_i may assume only the values 0 and 1, the structure function is

$$\phi(X_1, X_2, X_3, \ldots, X_n) = \prod_{i=1}^{n} X_i.$$

In the exercises, it is easily established that

$$\prod_{i=1}^{n} X_i = \min\{X_1, X_2, X_3, \ldots, X_n\}.$$

The 'minimum' form is often useful to describe the structure function for components in series.

By a **parallel system** we mean a configuration of n components as illustrated in Figure 3.2. A parallel system will function provided at least one (or more) of its components is functioning. It will therefore remain operational until the last component breaks. For a parallel system, the structure function is

$$\phi(X_1, X_2, X_3, \ldots, X_n) = 1 - \prod_{i=1}^{n}(1 - X_i).$$

This is easy to see because, on the left-side, $\phi = 1$ exactly when at least one of the Ci is operating, which occurs exactly when one or more of the X_i has the value 1, which means exactly that $\prod_{i=1}^{n}(1 - X_i) = 0$ or $1 - \prod_{i=1}^{n}(1 - X_i) = 1$. Therefore, the right- and left-sides agree when $\phi = 1$. Similar considerations for $\phi = 0$ yield the result. Note the useful alternative expressions:

$$1 - \prod_{i=1}^{n}(1 - X_i) = \max\{X_1, X_2, X_3, \ldots, X_n\}.$$

By a **k out of n system** we mean a system of n components which will function provided at least k (or more) of its components are functioning. This means that the structure function is

$$\phi(X_1, X_2, X_3, \ldots, X_n) = \begin{cases} 1 & \text{if } \sum_{i=1}^{n} X_i \geq k \\ 0 & \text{if } \sum_{i=1}^{n} X_i < k. \end{cases}$$

In a sense, this is a generalisation of the parallel system for which $k = 1$.

■ Example 3.1
Structure function of a simple computer network
In this example (Crowder et al., 1991) structure functions are constructed for simple computer systems such as the system specified in Figure 3.3.

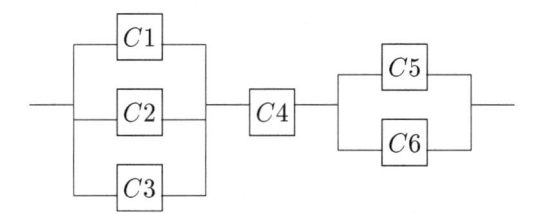

Figure 3.3 *A classical configuration of $n = 6$ components of which C1-C3 are terminals; C4 a computer; C5 and C6 local and central printers.*

The structure function is simply determined through max and min or sums and products of the indicator functions. First, the components $C1, C2$ and $C3$ are taken in parallel as one unit; next, $C4$ is taken as a single operational unit; then $C5$ and $C6$ are taken in parallel as a single operational unit; finally, the three units so constructed are linked in series. Therefore, the structure function is a product of parallel structure functions for the units:

$$\phi(X_1, X_2, X_3, \ldots, X_6)$$
$$= [1 - (1 - X_1)(1 - X_2)(1 - X_3)][X_4][1 - (1 - X_5)(1 - X_6)]$$
$$= \min\{\max\{X_1, X_2, X_3\}, X_4, \max\{X_5, X_6\}\}$$

□

We are primarily interested in cases where a component is important to the functioning of the system.

Definition 3.2 *The ith component Ci is **irrelevant** if, for all states of the other components in the system (that is, for all values of X_j for $j \neq i$),*

$$\phi(X_1, \ldots, X_{i-1}, 1, X_{i+1}, \ldots, X_n) = \phi(X_1, \ldots, X_{i-1}, 0, X_{i+1}, \ldots, X_n).$$

*A component is **relevant** when not irrelevant.*

□

We are also interested in systems where replacement of a component does not make the system 'worse off': that is, if a component is replaced so that the indicator value for the component changes from a zero to a one, the structure function either remains the same, or gets larger. That is, if we look at the position occupied by the ith component,

$$\phi(X_1, \ldots, X_{i-1}, 0, X_{i+1}, \ldots, X_n) \leq \phi(X_1, \ldots, X_{i-1}, 1, X_{i+1}, \ldots, X_n).$$

When we use the term **increasing structure function**, we mean that the structure function is non-decreasing in each argument. That is, **for each** i, and all states of the other components in the system,

$$\phi(X_1, \ldots, X_{i-1}, 0, X_{i+1}, \ldots, X_n) \leq \phi(X_1, \ldots, X_{i-1}, 1, X_{i+1}, \ldots, X_n). \quad (3.1)$$

We can express the concept of an increasing structure function using inequalities on the vector of indicators. We can write the state of the components in summary vector form by letting

$$\mathbf{X} = (X_1, X_2, \ldots, X_n)^T.$$

Then we use the notation $\mathbf{X} \leq \mathbf{Y}$ to mean that $X_i \leq Y_i$ with strict inequality for at least one $1 \leq i \leq n$. It follows from (3.1) that an increasing structure function satisfies

$$\mathbf{X} \leq \mathbf{Y} \quad \text{implies that} \quad \phi(\mathbf{X}) \leq \phi(\mathbf{Y}).$$

Definition 3.3 *A system is termed **coherent** if its structure function is increasing **and** each component is relevant.*

□

The structure function of a coherent system may be quite difficult to describe in simple terms. However, the following result indicates that the structure function of any coherent system is bounded above (and below) by the structure functions of parallel (and series) systems. Inevitably, this leads us to bounds on the reliability performance of coherent systems. The proof of Theorem 3.1 is considered in the exercises.

SIMPLE RELIABILITY

Theorem 3.1 *If ϕ is the structure function of a coherent system of n components in the state $\mathbf{X} = (X_1, X_2, \ldots, X_n)^T$, then*

$$\prod_{i=1}^{n} X_i \leq \phi(\mathbf{X}) \leq 1 - \prod_{i=1}^{n}(1 - X_i).$$

□

If we let $\phi_{series}(\mathbf{X})$ and $\phi_{parallel}(\mathbf{X})$ denote the structure functions for series and parallel alignments of the components in $\mathbf{X}$, then Theorem 3.1 implies that

$$\phi_{series}(\mathbf{X}) \leq \phi(\mathbf{X}) \leq \phi_{parallel}(\mathbf{X}).$$

3.2 Simple reliability

The probability that a component is operating (or its simple reliability) is easily determined from the indicator function representing the state of the component. By the **reliability of the ith component** with indicator function X_i, we mean the probability p_i given by

$$p_i = E(X_i).$$

Note that

$$E(X_i) = 0 \cdot P(X_i = 0) + 1 \cdot P(X_i = 1) = P(X_i = 1),$$

which is $P(Ci$ is functioning). It is convenient to use vector notation to collectively consider the the reliabilities of components in a system. As before, set $\mathbf{X} = (X_1, X_2, \ldots, X_n)^T$ as the vector of indicator functions and $\mathbf{p} = (p_1, p_2, p_3, \ldots, p_n)^T$ the vector of component reliabilities.

Definition 3.4 *The (simple)* **reliability** *of a system is given by the function h where $h(\mathbf{p}) = E[\phi(\mathbf{X})]$.*

□

In Definition 3.4, $h(\mathbf{p})$ represents the probability that the system is operating. Notice that

$$E[\phi(\mathbf{X})] = 1 \cdot P[\phi(\mathbf{X}) = 1] + 0 \cdot P[\phi(\mathbf{X}) = 0] = P[\phi(\mathbf{X}) = 1].$$

Therefore the probability that the system is operating is the expected value of the structure function. From this we can determine the reliability of coherent systems of independent components.

3.3 Independent components

When components operate independently, it means that the indicator functions $X_1, X_2, X_3, \ldots, X_n$ representing the components are mutually independent as random variables. Intuitively, whether or not component j is functioning does not influence the functioning of another component i.

Theorem 3.2 *The reliability of a system of components with indicators* $\mathbf{x}$ *and structure function ϕ is given by* $h(\mathbf{p}) = \sum_{\mathbf{x}} \phi(\mathbf{x}) \prod_{i=1}^{n} [p_i^{x_i} (1-p_i)^{1-x_i}]$.

Proof:

$$\begin{aligned}
E[\phi(\mathbf{X})] &= \sum_{\mathbf{x}} \phi(\mathbf{x}) P(\mathbf{X} = \mathbf{x}) \\
&= \sum_{\mathbf{x}} \phi(\mathbf{x}) \prod_{i=1}^{n} P(X_i = x_i) \quad \text{(by independence)} \\
&= \sum_{\mathbf{x}} \phi(\mathbf{x}) \prod_{i=1}^{n} [p_i^{x_i} (1-p_i)^{1-x_i}].
\end{aligned}$$

□

In the expression for $h(\mathbf{p})$ in Theorem 3.2, the terms $\phi(\mathbf{X})$ assume either the value 0 or 1 as $\mathbf{X}$ changes. Therefore the reliability is a sum of products and powers of p_i and $(1-p_i)$. In Example 3.2, we can see simple instances of this formulation.

■ **Example 3.2**
Reliability of series, parallel, k out of n systems
Consider the following instances of Theorem 3.2 for n independent components with reliability vector $\mathbf{p} = (p_1, p_2, p_3, \ldots, p_n)^T$.

Series system: product law of reliability

$$h(\mathbf{p}) = \prod_{i=1}^{n} p_i$$

Parallel system: product law of unreliability

$$h(\mathbf{p}) = 1 - \prod_{i=1}^{n} (1 - p_i)$$

k **out of** n **system:** for the case $p = p_1 = p_2 = p_3 = \ldots = p_n$,

$$h(\mathbf{p}) = \sum_{j=k}^{n} \binom{n}{j} p^j (1-p)^{n-j}$$

For example, the case of independent parallel components could be fully expressed as

$$\begin{aligned}
h(\mathbf{p}) &= P(\max\{X_1, X_2, \ldots, X_n\} = 1) \\
&= P(\text{at least one } X_i \text{ equals } 1) \\
&= 1 - P(\text{all } X_i \text{ are zero}) \\
&= 1 - P(X_1 = 0, X_2 = 0, \ldots, X_n = 0) \\
&= 1 - P(X_1 = 0) P(X_2 = 0) \ldots P(X_n = 0) \\
&= 1 - (1-p_1)(1-p_2) \ldots (1-p_n),
\end{aligned}$$

or, even from the point of the expected value of the structure function as:

$$\begin{aligned} h(\mathbf{p}) &= E[\phi(\mathbf{X})] \\ &= E(\max\{X_1, X_2, \ldots, X_n\}) \\ &= E[1 - \prod_{i=1}^{n}(1 - X_i)] \\ &= 1 - \prod_{i=1}^{n}[1 - E(X_i)] \\ &= 1 - (1 - p_1)(1 - p_2)\ldots(1 - p_n). \end{aligned}$$

$\square$

3.4 Paths and cuts

The reliability of complex systems is difficult to write down using the statement of Theorem 3.2. The series and parallel cases listed in the previous example are easily stated and, in fact, form the building blocks for estimating the reliability of more complex systems. This is done through sets of **paths** and **cuts**.

Definition 3.5 *A* **path** *is a subsystem of components with the property that if all components on the path are functioning then the system is functioning. A* **cut** *is a subsystem of components with the property that if all components on the cut fail then the system fails. A path (cut) is* **minimal** *if no strict subset has the property of being a path (cut).* $\square$

If we list the minimal paths as $P_1, P_2, \ldots, P_p$ and the minimal cuts as $K_1, K_2, \ldots, K_k$, then, since all components on at least one path must function for the system to function, we may write the structure function as

$$\phi(\mathbf{X}) = \max_{1 \leq j \leq p} \min_{i \in P_j} X_i. \tag{3.2}$$

Similarly, if at least one component in every cut functions then the system will function and we may write the structure function as

$$\phi(\mathbf{X}) = \min_{1 \leq j \leq k} \max_{i \in K_j} X_i. \tag{3.3}$$

Such representations of the structure function are illustrated in the next example.

■ Example 3.3
Minimal paths and cuts of the computer system in Example 3.1
The computer system of Example 3.1 has structure function

$$\phi(X_1, X_2, X_3, \ldots, X_6) = \min\{\max\{X_1, X_2, X_3\}, X_4, \max\{X_5, X_6\}\}.$$

This corresponds to minimal path sets: $P_1 = \{1,4,5\}, P_2 = \{2,4,5\}, P_3 = \{3,4,5\}, P_4 = \{1,4,6\}, P_5 = \{2,4,6\}, P_6 = \{3,4,6\}$ and minimal cut sets $K_1 = \{1,2,3\}, K_2 = \{4\}, K_3 = \{5,6\}$. Notice that the cut sets are disjoint and the reliability of the system is obtained by placing parallel arrangements of each cut set in a series. □

The reliability is easily determined by (3.3) as the expected value of the minimum of a set of independent random variables $\max_{i \in K_j} X_i$ which are independent when the minimal cut sets are disjoint. Thus:

$$h(\mathbf{p}) = E[\phi(\mathbf{X})] = \prod_{1 \leq j \leq k} E[\max_{i \in K_j} X_i] = \prod_{1 \leq j \leq k} [1 - \prod_{i \in K_j} (1 - p_i)]. \quad (3.4)$$

For the simple computer system under investigation since Example 3.1, the minimal cut set representation of the reliability in (3.4) is

$$h(\mathbf{p}) = [1 - (1 - p_1)(1 - p_2)(1 - p_3)][p_4][1 - (1 - p_5)(1 - p_6)].$$

In general, however, the determination of $h(\mathbf{p}) = E[\phi(\mathbf{X})]$ from (3.2) and (3.3) involves taking the expected value across a dependent set of random variables, making explicit expressions for reliability difficult. Bounds on the the reliability may be obtained through the concept of positive **association**. This is based on the covariance between increasing functions of the vector of indicator variables.

3.5 Associated components

According to Barlow and Proschan (1975), lack of independence between components, or groupings of components, can occur when components are subjected to the same set of stresses, share load, or are common to more than one set of paths or cuts. Even though it may be unrealistic to assume that the components are independent in these circumstances, they may, however, be considered 'associated'. Essentially this means that the random variables tend to behave similarly: if a coherent system is stressed, then all the components tend to be similarly adversely affected; if a component fails, then this will adversely affect all minimal path sets containing that component. We give the definition of associated for a vector of indicator functions.

Definition 3.6 *Binary 0-1 variables* $\mathbf{Z} = (Z_1, Z_2, \ldots, Z_r)^T$ *are* **associated** *if* $Cov[f(\mathbf{Z}), g(\mathbf{Z})] \geq 0$, *for all pairs of increasing binary functions* f *and* g.
□

It becomes useful to define the **dual**, f^D, of a binary function f by

$$f^D(\mathbf{Z}) = 1 - f(\mathbf{1} - \mathbf{Z}). \quad (3.5)$$

There is a natural interpretation of the dual. The dual of the structure function for a series system is the structure function for a parallel system and vice-versa.

ASSOCIATED COMPONENTS

By (3.5), if f is binary increasing, then f^D is also binary increasing. Using the dual, it follows that if the binary variables in $\mathbf{Z}$ are associated, then so are the binary variables in $\mathbf{1} - \mathbf{Z}$.

Theorem 3.3 is our first insight into the usefulness of associated components. It shows that if we assume components to be independent when in fact they are associated but not independent, the system reliability will be underestimated. To see this, suppose that the binary variables in the r entries in the vector $\mathbf{Z}$ are associated. In particular, the functions Z_1 and $\prod_{i=2}^{r} Z_i$ are examples of a pair of increasing functions of $\mathbf{Z}$. (You should be able to check that this is so.) By definition of the term associated,

$$Cov[Z_1, \prod_{i=2}^{r} Z_i] \geq 0.$$

Theorem 3.3 follows by repeated use of such covariance expressions.

Theorem 3.3 *If* $\mathbf{Z} = (Z_1, Z_2, \ldots, Z_r)^T$ *are associated binary random variables, then* $P(Z_1 = 1, Z_2 = 1, \ldots, Z_r = 1) \geq P(Z_1 = 1) \ldots P(Z_r = 1)$. □

It is interesting to note that if $\mathbf{Z}$ are associated, then $\mathbf{1} - \mathbf{Z}$ are associated. This means that Theorem 3.3 applies to the variables $\mathbf{Z}$ being replaced by the variables $\mathbf{1} - \mathbf{Z}$. This means that:

$$P(Z_1 = 0, Z_2 = 0, \ldots, Z_r = 0) \geq P(Z_1 = 0) \ldots P(Z_r = 0). \quad (3.6)$$

There are several important uses of Theorem 3.3. An immediate and important use of is to establish that if ϕ is a coherent structure of associated components with reliability $\mathbf{p} = (p_1, p_2, \ldots, p_n)^T$, then by taking expected values throughout the inequality in Theorem 3.1, it follows that

$$\prod_{i=1}^{n} p_i \leq h(\mathbf{p}) \leq 1 - \prod_{i=1}^{n} (1 - p_i). \quad (3.7)$$

This is our first useful inequality on performance bounds for systems of associated components. We can obtain improvements to these bounds by examining the interactions of the path sets and cut sets.

Suppose that the component indicator functions $\mathbf{X} = (X_1, X_2, \ldots, X_n)^T$ are associated. Define in terms of $\mathbf{X}$ the **minimal path binary indicators**

$$Z_j(\mathbf{X}) = \min_{i \in P_j} X_i, \quad j = 1, 2, 3, \ldots, p,$$

which are increasing functions of associated binary variables and hence associated. Using (3.6), it follows that

$$\begin{aligned} h(\mathbf{p}) &= P(\phi(\mathbf{X}) = 1) \\ &= 1 - P(Z_1(\mathbf{X}) = 0, Z_2(\mathbf{X}) = 0, \ldots, Z_p(\mathbf{X}) = 0) \end{aligned}$$

$$\leq 1 - \prod_{j=1}^{p} P(Z_j(\mathbf{X}) = 0)$$

$$= 1 - \prod_{j=1}^{p} [1 - P(Z_j(\mathbf{X}) = 1)].$$

Similarly, applying this Theorem 3.3 to **minimal cut structure functions**

$$W_j(\mathbf{X}) = \max_{i \in K_j} X_i, \quad j = 1, 2, 3, \ldots k,$$

gives

$$\begin{aligned} h(\mathbf{p}) &= P(\phi(\mathbf{X}) = 1) \\ &= P(W_1(\mathbf{X}) = 1, W_2(\mathbf{X}) = 1, \ldots, W_k(\mathbf{X}) = 1) \\ &\geq \prod_{j=1}^{k} P(W_j(\mathbf{X}) = 1). \end{aligned}$$

This gives the new set of reliability bounds for associated components:

$$\prod_{j=1}^{k} P(W_j(\mathbf{X}) = 1) \leq h(\mathbf{p}) \leq 1 - \prod_{j=1}^{p} [1 - P(Z_j(\mathbf{X}) = 1)].$$

Further simplification is possible when the components are not only associated, but operate independently. This is because independence is a stronger assumption than association (Barlow and Proschan, 1975). $P(W_j(\mathbf{X}) = 1)$ and $P(Z_j(\mathbf{X}) = 1)$ both factorise when the variables in $\mathbf{X}$ are independent. (See Example 3.4.)

This final independence assumption yields lower (upper) bounds for $h(\mathbf{p})$ by replacing each component in the cut (path) sets by a corresponding set of independently operating components of the same reliability and then assuming all components in the whole structure to be independent. Thus:

Theorem 3.4 *Consider a coherent system of independent components with structure function ϕ. The reliability function $h(\mathbf{p})$ satisfies*

$$\prod_{1 \leq j \leq k} [1 - \prod_{i \in K_j} (1 - p_i)] \leq h(\mathbf{p}) \leq 1 - \prod_{1 \leq j \leq p} [1 - \prod_{i \in P_j} p_i].$$

□

We illustrate Theorem 3.4 by determining the reliability bounds for a single bridge structure.

■ Example 3.4
Reliability bounds for a single bridge structure

Consider the system in Figure 3.4 where energy flows, as usual, from left to right. Component 3 corresponds to a bridge structure.

TIME-DEPENDENT RELIABILITY

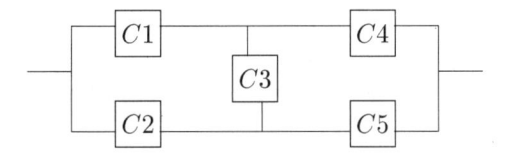

Figure 3.4 *A configuration of components with a bridge structure.*

For this system, the minimal path sets are $P_1 = \{1,4,\}, P_2 = \{2,5\}, P_3 = \{1,3,5\}, P_6 = \{2,3,4\}$ and minimal cut sets $K_1 = \{1,2\}, K_2 = \{4,5\}, K_3 = \{1,3,5\}, K_3 = \{2,3,4\}$. In this case, groups of components are not operating independently since some sets are both path and cut sets, and the cut sets are not disjoint (as was the case for the computer system of Example 3.1.) If we assume that individual components are operating independently, then they are associated and Theorem 3.4 provides an appropriate bound for the reliability.

To illustrate the upper bound in Theorem 3.4, we connect each path set in a series and then link these series systems in parallel. Using Theorem 3.3,

$$\begin{aligned} h(\mathbf{p}) &= 1 - P(\text{all paths fail}) \\ &= 1 - P(X_1 X_4 = 0, X_2 X_5 = 0, X_1 X_3 X_5 = 0, X_2 X_3 X_4 = 0) \\ &\leq 1 - P(X_1 X_4 = 0) P(X_2 X_5 = 0) P(X_1 X_3 X_5 = 0) P(X_2 X_3 X_4 = 0) \\ &= 1 - (1 - p_1 p_4)(1 - p_2 p_5)(1 - p_1 p_3 p_5)(1 - p_2 p_3 p_4). \end{aligned}$$

Then we connect each cut set in parallel and link these parallel systems in series.

$$\begin{aligned} h(\mathbf{p}) &= P(\text{at least one of each cut set is operational}) \\ &= P[(1 - X_1)(1 - X_2) = 0, (1 - X_4)(1 - X_5) = 0, \\ &\quad (1 - X_1)(1 - X_3)(1 - X_5) = 0, (1 - X_2)(1 - X_3)(1 - X_4) = 0] \\ &\geq P[(1 - X_1)(1 - X_2) = 0] P[(1 - X_4)(1 - X_5) = 0] \\ &\quad P[(1 - X_1)(1 - X_3)(1 - X_5) = 0] P[(1 - X_2)(1 - X_3)(1 - X_4) = 0] \\ &= [1 - (1 - p_1)(1 - p_2)][1 - (1 - p_4)(1 - p_5)] \\ &\quad [1 - (1 - p_1)(1 - p_3)(1 - p_5)][1 - (1 - p_2)(1 - p_3)(1 - p_4)] \end{aligned}$$

□

3.6 Time-dependent reliability

We begin by examining series and parallel component failure when the components are operating independently. Suppose that the independent random variables $Y_1, Y_2, Y_3, \ldots, Y_n$ give the lifetimes (or times to failure) of n components, where Y_i is the time to failure of Ci, $i = 1, 2, \ldots, n$. According to

Definition 1.2, the **reliability** of Ci at time y is
$$S_i(y) = P(Y_i \geq y).$$
The main result of this section is the general statement of the reliability of the series and parallel systems.

Theorem 3.5 *(Product Law of Reliability)* *When components Ci with reliabilities S_i, $i = 1, 2, \ldots, n$, are connected in series under independent operation, the reliability S at time y of the resulting system is*
$$S(y) = S_1(y)S_2(y)\ldots S_n(y).$$

(Product Law of Unreliability) When the components are connected in parallel under independent operation, the reliability S at time y of the resulting system is
$$1 - S(y) = [1 - S_1(y)][1 - S_2(y)]\ldots[1 - S_n(y)].$$
□

We establish these reliability laws for the useful special case where the components are of the same type, and that their lifetimes follow exponential distributions with common mean β. Thus, for each i, the random variables Y_i each have $exp(\beta)$ distributions with common reliability
$$S_1(y) = S_2(y) = \ldots = S_n(y) = e^{-\frac{y}{\beta}}.$$

If the components are all connected in series, then the resulting series system will operate precisely when all the components are operating. That is, the system fails to operate when the first component fails. Therefore, the random variable, Y_{series}, describing the lifetime of the entire series system is the minimum of all the possible lifetimes $Y_1, Y_2, Y_3, \ldots, Y_n$. To determine the distribution of
$$Y = Y_{series} = Y_{(1)} = \min\{Y_1, Y_2, Y_3, \ldots, Y_n\},$$
we may use the cumulative distribution function method, which identifies a distribution through the structure of its cdf.

$$\begin{aligned}
F_Y(y) &= P(Y \leq y) \\
&= 1 - P(Y > y) \\
&= 1 - P(Y_{(1)} > y) \\
&= 1 - P(\min\{Y_1, Y_2, Y_3, \ldots, Y_n\} > y) \\
&= 1 - P(Y_1 > y \text{ and } Y_2 > y \text{ and } \ldots, \text{ and } Y_n > y) \\
&= 1 - P(Y_1 > y, Y_2 > y, \ldots, Y_n > y) \\
&= 1 - P(Y_1 > y)P(Y_2 > y)\ldots P(Y_n > y) \quad \text{(by independence)} \\
&= 1 - (e^{-\frac{y}{\beta}})(e^{-\frac{y}{\beta}})\ldots(e^{-\frac{y}{\beta}}) \\
&= 1 - e^{-n\frac{y}{\beta}}.
\end{aligned}$$

Hence, $Y_{series} \sim exp(\frac{\beta}{n})$. Since $E(Y_{series}) = \frac{\beta}{n}$, it follows that the mean lifetime has been *reduced* by a factor of $\frac{1}{n}$ by having the components connected in series. The reliability of this series system is $S(y) = e^{-n\frac{y}{\beta}}$. This helps explain why a large number of lights connected in series on a Christmas tree may fail very rapidly. We emphasise the importance of this result by restating it as a theorem:

Theorem 3.6 *If survival random variables* $Y_1, Y_2, Y_3, \ldots, Y_n$ *are i.i.d.* $exp(\beta)$, *then* $Y_{(1)} = \min\{Y_1, Y_2, \ldots Y_n\} \sim exp(\frac{\beta}{n})$. □

An interesting result occurs if the components individually have different survival distributions. For each $i = 1, 2, \ldots, n$, suppose that the random variables Y_i each have $exp(\beta_i)$ distributions with reliability

$$S_i(y) = e^{-\frac{y}{\beta_i}}.$$

Then, as before,

$$\begin{aligned}
F_Y(y) &= P(Y \leq y) \\
&= 1 - P(Y_1 > y)P(Y_2 > y)\ldots P(Y_n > y) \quad \text{(by independence)} \\
&= 1 - (e^{-\frac{y}{\beta_1}})(e^{-\frac{y}{\beta_2}})\ldots(e^{-\frac{y}{\beta_n}}) \\
&= 1 - e^{-y(\frac{1}{\beta_1} + \frac{1}{\beta_2} + \ldots + \frac{1}{\beta_n})} \\
&= 1 - e^{-n(\frac{y}{\beta^*})},
\end{aligned}$$

where

$$\beta^* = \left(\frac{\frac{1}{\beta_1} + \frac{1}{\beta_2} + \ldots + \frac{1}{\beta_n}}{n}\right)^{-1}$$

is the **harmonic mean** of $\beta_1, \beta_2, \ldots, \beta_n$.

If the components are connected in parallel, then the resulting system will operate precisely when at least one of the components is operating. That is, the system fails to operate when the longest-running component fails. Therefore, the random variable, $Y_{parallel}$, describing the lifetime of the entire parallel system is the maximum $Y_{(n)}$ of all the possible lifetimes $Y_1, Y_2, Y_3, \ldots, Y_n$. Then,

$$\begin{aligned}
F_Y(y) &= P(Y \leq y) \\
&= P(Y_{(n)} \leq y) \\
&= P(\max\{Y_1, Y_2, Y_3, \ldots, Y_n\} \leq y) \\
&= P(Y_1 \leq y \text{ and } Y_2 \leq y \text{ and } \ldots, \text{ and } Y_n \leq y) \\
&= P(Y_1 \leq y, Y_2 \leq y, \ldots, Y_n \leq y) \\
&= P(Y_1 \leq y)P(Y_2 \leq y)\ldots P(Y_n \leq y) \quad \text{(by independence)} \\
&= (1 - e^{-\frac{y}{\beta}})(1 - e^{-\frac{y}{\beta}})\ldots(1 - e^{-\frac{y}{\beta}}) \\
&= (1 - e^{-\frac{y}{\beta}})^n.
\end{aligned}$$

The probability density function for $Y_{parallel}$ may be found by differentiation.

Such a density is difficult to integrate when trying to find the MTTF for a parallel system, which is more successfully determined by Theorem 3.7.

Theorem 3.7 *If $Y_1, Y_2, Y_3, \ldots, Y_n$ are i.i.d. $exp(\beta)$, and $Y = Y_{parallel} = Y_{(n)} = \max\{Y_1, Y_2, Y_3, \ldots, Y_n\}$, then*

$$E[Y] = \beta \left[1 + \frac{1}{2} + \frac{1}{3} + \frac{1}{4} + \ldots + \frac{1}{n-1} + \frac{1}{n} \right].$$

Proof: The following diagram illustrates how a parallel system fails at $y_{(n)}$.

Lifetime of parallel system

According to Theorem 3.6, the mean time for $D_1 = Y_{(1)}$ to occur is $\frac{\beta}{n}$ when counting of time starts from 0. Because of the memoryless nature of the exponential model, the system continues to operate, as before, but now with one less component. Therefore, the time $D_2 = Y_{(2)} - Y_{(1)}$ for the second component to break has an $exp(\frac{\beta}{n-1})$ distribution. Similarly, the time $D_3 = Y_{(3)} - Y_{(2)}$ for the second component to break has an $exp(\frac{\beta}{n-2})$ distribution, ... the time $D_n = Y_{(n)} - Y_{(n-1)}$ for the last component to break (and the parallel system to fail) has an $exp(\beta)$ distribution. The difference random variables $D_1, D_2, D_3, \ldots, D_n$ are independent, and the result follows by taking the expected value of each side of the equation $D_1 + D_2 + D_3 + \ldots + D_n = Y_{(n)}$. □

Theorem 3.7 shows that there are diminishing returns from connecting a large number of components in parallel.

3.7 Exercises

3.1. Show that the indicator functions $X_1, X_2, X_3, \ldots, X_n$ for a system of components $C1, C2, C3, \ldots Cn$ satisfy

$$\min\{X_1, X_2, X_3, \ldots, X_n\} = \prod_{i=1}^{n} X_i$$

and $\max\{X_1, X_2, X_3, \ldots, X_n\} = 1 - \prod_{i=1}^{n}(1 - X_i)$.

3.2. Consider identical components $C1, C2, C3, C4, C5$ operating independently. System i, labelled K_i, is constructed as:

K_1 consists of $C1$ and $C2$ in series;
K_2 consists of $C3$ and $C4$ in parallel;
K_3 consists of $C5$; and
K consists of K_1, K_2 and K_3 in series.

(a) Draw a diagram of the system K.
(b) Find the minimal path and minimal cut sets for K.

EXERCISES

(c) Give two forms for the structure function of K and show that they are equivalent.

(d) Determine the reliability of K, K_1, K_2 and K_3 if $p_i = P(Ci \text{ operates})$.

3.3. Consider the following notation for a system of n components with indicator variables $X_1, X_2, \ldots, X_n$:

$$(1_i, \mathbf{X}) \equiv (X_1, \ldots, X_{i-1}, 1, X_{i+1}, \ldots, X_n)$$
$$(0_i, \mathbf{X}) \equiv (X_1, \ldots, X_{i-1}, 0, X_{i+1}, \ldots, X_n)$$

Show that the following identity holds for any structure function ϕ of the n components:

$$\phi(\mathbf{X}) = X_i \phi(1_i, \mathbf{X}) + (1 - X_i)\phi(0_i, \mathbf{X})$$

for all $\mathbf{X}$ ($i = 1, 2, \ldots, n$).

3.4. Show that, for a coherent system,

$$\phi(1, 1, 1, \ldots, 1) = 1 \quad \text{and} \quad \phi(0, 0, 0, \ldots, 0) = 0.$$

3.5. If ϕ is a coherent structure function of n components, show that

$$\prod_{i=1}^{n} X_i \leq \phi(\mathbf{X}) \leq 1 - \prod_{i=1}^{n}(1 - X_i).$$

What types of systems do the endpoints of these inequalities correspond to?

3.6. Suppose that two electronic components have lifetimes Y_1 and Y_2 that are independent random variables with

$$Y_1 \sim exp(\lambda_1) \text{ and } Y_2 \sim exp(\lambda_2).$$

Determine $P(Y_1 > Y_2)$. What occurs when $\lambda_1 = \lambda_2$?

3.7. An electronic unit consists of two components, $C1$ and $C2$, arranged in parallel. Suppose that Y_1 denotes the lifetime of $C1$ and Y_2 denotes the lifetime of $C2$. Assume that Y_1 and Y_2 are independent.

(a) If Y_1, Y_2 are i.i.d. $exp(\lambda)$ random variables, determine the probability density function of the lifetime of the unit.

(b) If $Y_i \sim exp(\lambda_i)$, $i = 1, 2$, determine the probability density function of the lifetime of the unit.

(c) If Y_1, Y_2 are i.i.d. $exp(\lambda)$ random variables and two such units are connected in series, determine the probability density function of the lifetime of the resulting system.

(d) Discuss the above results applied to n components.

3.8. An electronic unit consists of two components, $C1$ and $C2$, arranged in series.

Suppose that Y_1 denotes the lifetime of $C1$ and Y_2 denotes the lifetime of $C2$. Assume that Y_1 and Y_2 are independent.

(a) If Y_1, Y_2 are i.i.d. $exp(\lambda)$ random variables, determine the probability density function of the lifetime of the unit.

(b) If $Y_i \sim exp(\lambda_i)$, $i = 1, 2$, determine the probability density function of the lifetime of the unit.

(c) If Y_1, Y_2 are i.i.d. $exp(\lambda)$ random variables and two such units are connected in parallel, determine the probability density function of the lifetime of the resulting system.

(d) Discuss the above results applied to n components.

3.9. Components $C1$ and $C2$ possess reliability functions

$$S_1(t) = \begin{cases} -\frac{t}{2} + 1, & \text{if } 0 \le t \le 2 \\ 0, & \text{otherwise} \end{cases}$$

and

$$S_2(t) = \begin{cases} -2t + 1, & \text{if } 0 \le t \le \frac{1}{2} \\ 0, & \text{otherwise,} \end{cases}$$

respectively.

(a) Based on the mean time to failure (MTTF), which system is more reliable?

(b) Determine the MTTF when the two systems are connected in series.

(c) Determine the MTTF when the two systems are connected in parallel.

3.10. Consider the following two different computer systems:

System 1: Two memory components in parallel — $C1$; two CPU components in parallel — $C2$; then $C1$ and $C2$ combined in series.

System 2: A memory component and a CPU component in series — $C1$; a memory component and a CPU component in series — $C2$; then $C1$ and $C2$ combined in parallel.

Determine which system appears to be the most reliable based on a MTTF criterion when component failures are assumed to be independent and each component (memory and CPU) is assumed to possess the reliability function

$$S_c(t) = \begin{cases} 1 - \frac{t}{3}, & \text{if } 0 \le t \le 3 \\ 0, & \text{otherwise.} \end{cases}$$

3.11. Suppose that an aeroplane has four engines which operate independently. The lifetime of each engine is exponentially distributed with mean 1000 hours.

EXERCISES

(a) Determine the hazard function for the time, Y, until the first engine fails. Also, determine the mean and standard deviation of Y.

(b) The aeroplane can remain in flight if at least two of its engines are working. Determine the mean and standard deviation of the time until the plane is forced to land with engine problems.

3.12. Suppose that for each $i = 1, 2, 3, 4, 5$, System i consists of 3 components in series with each component having constant hazard $h(y) = 20$. Suppose that a Unit consists of Systems 1 to 5 connected in parallel. Determine the mean time to failure of the Unit.

3.13. Components $C1, C2, \ldots, Cn$ operate independently in a system. Suppose that for each $i = 1, 2, \ldots, n$, Y_i, the time to failure for Ci, follows an exponential distribution for which $Y_i \sim exp(\beta)$.

Let $Y_{parallel}$, the survival random variable (time to failure) of the system when the components are linked in parallel, have hazard function $h_{parallel}(y)$.

Let Y_{series}, the survival random variable (time to failure) of the system when the components are linked in series, have hazard function $h_{series}(y)$.

Further suppose that a function $B_n(y)$ is defined by

$$B_n(y) = \frac{e^{-\frac{y}{\beta}}(1 - e^{-\frac{y}{\beta}})^{n-1}}{1 - (1 - e^{-\frac{y}{\beta}})^n}.$$

(a) Derive an expression (in terms of y and β) for the survival function of the series system.

(b) Derive an expression for $h_{series}(y)$ in terms of y and β.

(c) Show that
$$h_{parallel}(y) = h_{series}(y) B_n(y).$$

(d) Show that $B_n(y)$ satisfies:

(i.) $B_n(0) = 0$;

(ii.) $B_n(y) \leq 1$ for all y; [Hint: Evaluate $B_n(y)$ with the substitution $u = 1 - e^{-\frac{y}{\beta}}$ and assume the fact that $1 - u^n$ has $1 - u$ as a factor.] What is the interpretation of $B_n(y) \leq 1$ in terms of series and parallel systems?

(iii.) $B_n(y) \to \frac{1}{n}$ as $y \to \infty$. Interpret this result in terms of a parallel system where the number of components is large.

3.14. Show that if $\mathbf{Z} = (Z_1, Z_2, \ldots, Z_r)^T$ are associated binary random variables, then

$$P(Z_1 = 1, Z_2 = 1, \ldots, Z_r = 1) \geq P(Z_1 = 1)P(Z_2 = 1) \ldots P(Z_r = 1).$$

[Hint: Consider $Cov(Z_1, Z_2 Z_3 \ldots Z_r)$.]

3.15. Show that if f is a binary increasing function then f^D, the dual of f defined by $f^D(\mathbf{Z}) = 1 - f(\mathbf{1} - \mathbf{Z})$ is also binary increasing.

3.16. Show that if $\mathbf{Z} = (Z_1, Z_2, \ldots, Z_r)^T$ are associated binary random variables, then $\mathbf{1} - \mathbf{Z} = (1 - Z_1, 1 - Z_2, \ldots, 1 - Z_r)^T$ are associated binary random variables. Use this result, along with

$$P(Z_1 = 1, Z_2 = 1, \ldots, Z_r = 1) \geq P(Z_1 = 1)P(Z_2 = 1) \ldots P(Z_r = 1),$$

to prove that

$$P(Z_1 = 0, Z_2 = 0, \ldots, Z_r = 0) \geq P(Z_1 = 0)P(Z_2 = 0) \ldots P(Z_r = 0).$$

3.17. The hazard rate for the guidance system for on-board control of a space vehicle is thought to follow the following power function of time:

$$h(t) = \alpha \mu t^{\alpha - 1} + \beta \gamma t^{\beta - 1}, \quad \alpha > 0, \beta > 0, \mu > 0 \text{ and } \gamma > 0.$$

Determine the reliability of the system. Does the system appear to have an underlying structure of two separate components?

3.18. Let $Y_1, Y_2, \ldots, Y_n$ denote a random sample from an exponential distribution with mean β. If F_n is the distribution function of the *normalised* random variable

$$\frac{Y_{(n)} - a_n}{b_n} \quad \text{for} \quad Y_{(n)} = \max\{Y_1, Y_2, \ldots, Y_n\},$$

where $a_n = \beta \log_e n$ and $b_n = \beta$, show that

$$\lim_{n \to \infty} F_n(y) = \exp(-e^{-y}).$$

CHAPTER 4

Data Plots

Suppose that we have obtained data on the survival random variable Y. How can we use the data to assess a likely probability model for Y? In this chapter we discover that QQ-plots and hazard plots can be of assistance.

However, if we begin by not assuming a model for the distribution for Y, but rather listen to what the data suggests for the shape of the survival function and the hazard function of Y, what techniques are then available? A very simple technique for looking at the distributional shape implied by data is the **empirical survivor function** defined in Chapter 1 and now discussed further in Section 4.1.

Recall that observed survival times may be written $y_1, y_2, y_3, \ldots, y_n$ in the order in which the observations are recorded. The notation indicates that the data set has n data points. The notation for the **ranked data set**, which represents $y_1, y_2, y_3, \ldots, y_n$ arranged in rank order from smallest to largest, is

$$y_{(1)}, y_{(2)}, y_{(3)}, \ldots y_{(n-1)}, y_{(n)},$$

for which $y_{(1)} \leq y_{(2)} \leq y_{(3)} \leq \cdots \leq y_{(n-1)} \leq y_{(n)}$. This notation will be particularly helpful in describing data plots and how their axes are labelled.

4.1 Empirical survivor function

In Section 1.3 it was demonstrated how to estimate the survival function $S(y) = P(Y > y)$ in a distribution-free way at specified values of y using the empirical survivor function

$$S_n(y) = \frac{\text{number of observations} > y}{n} = \frac{1}{n} \sum_{i=1}^{n} I_{(y,\infty)}(Y_i) \qquad (4.1)$$

and is an estimate of the survival function $S(y) = P(Y > y)$ based on observations $Y_1, Y_2, Y_3, \ldots, Y_n$. To get an estimation of the accuracy of the estimation process, it was further shown that at a fixed point $y = y^*$, we estimate $P(Y > y^*)$, the probability of survival beyond y^* as the approximate confidence interval based on two (estimated) standard errors:

$$S_n(y^*) \pm 2\widehat{S.E.}[S_n(y^*)],$$

where

$$\widehat{S.E.}[S_n(y^*)] = \sqrt{\frac{S_n(y^*)[1 - S_n(y^*)]}{n}}.$$

Consider now how a graph of $S_n(y)$ from (4.1) may be constructed as a function of y. Such a graph should provide insight into the shape of the true survival function $S(y)$ as a function of y. Clearly, the function $S_n(y)$ has initial value 1, as all data lie beyond zero. Then, as y increases, the function remains constant until the first (smallest) data point is reached; $S_n(y)$ then reduces in size by $\frac{1}{n}$, again remaining constant until the second data point $y_{(2)}$ is reached; again a reduction by $\frac{1}{n}$, and so on. Therefore, the function is piecewise linear with jumps of size $\frac{1}{n}$ at each of the n data points. Of course, if j of the data are tied (that is, have the same value), then the jump size is collectively $\frac{j}{n}$ at the tied point. Finally, the function reaches 0 at $y = y_{(n)}$. The start and finish of such a graph are illustrated in Figure 4.1, where the the right-continuity of the function is shown through a sequence of hollow and filled in dots.

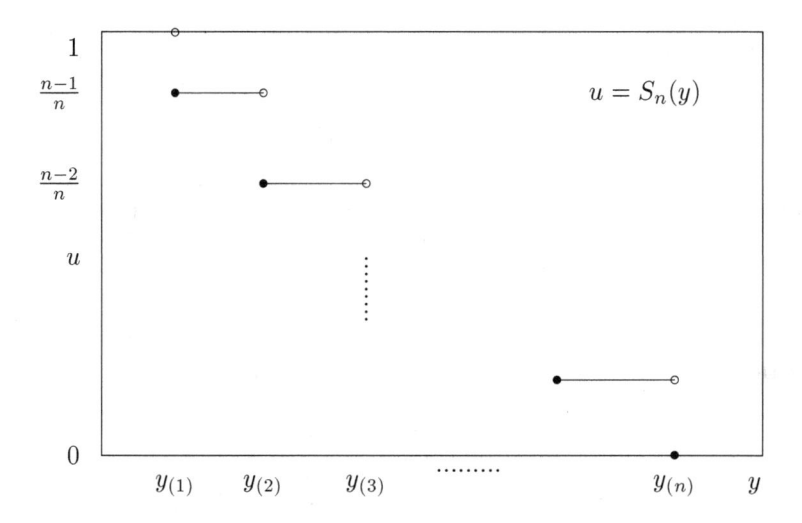

Figure 4.1 *Graph of the empirical survivor function $u = S_n(y)$ for a survival random variable Y based on a data set of n distinct points.*

The graph in Figure 4.1 is difficult to draw using software in a way that shows both the right-continuity and the piecewise linearity. A simple alternative that has enormous benefits in showing the estimated shape of the survival curve is obtained by plotting points located in the middle of the 'jumps' on the vertical axis and at the observed data points on the horizontal axis. To define such a plot we need to develop the notion of 'plotting positions' in the next section.

4.2 Sample quantile function

Figure 4.1 shows how difficult it is to develop a one-to-one relationship between the ranked plotted data on the horizontal axis and the $[0, 1]$ scale on the

SAMPLE QUANTILE FUNCTION

vertical axis. For example, if we chose a particular fraction between 0 and 1 on the vertical axis in Figure 4.1 and traced it back to the horizontal axis, what horizontal axis point would we choose? That is, given a particular fraction, what data point has this fraction of data points smaller than itself? To answer this we seek the sample quantiles.

Following Parzen (1979), the **sample quantile function** may be defined in terms of the empirical survivor function S_n.

Definition 4.1 *For $0 \leq u \leq 1$, the **sample quantile function** $Q_n(u)$ is* $Q_n(u) = \inf\{y : S_n(y) \leq 1 - u\}$. □

This definition makes Q_n the inverse function to $\underline{F} = 1 - S_n$; that is, $Q_n(u) = S_n^{-1}(1-u) = \underline{F}^{-1}(u)$ for $\underline{F} = 1 - S_n$. See, for example, Parzen (1979) for an extensive discussion of the properties of Q_n. The construction of Q_n as an inverse function is achieved by tracing the dotted line in Figure 4.2.

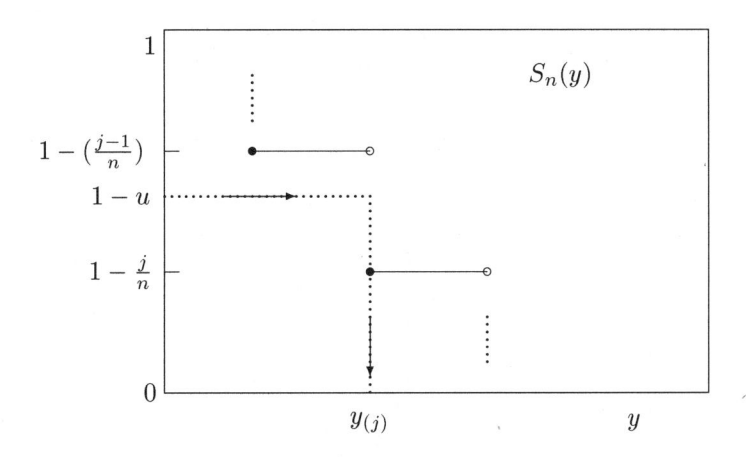

Figure 4.2 *Graph of the empirical survivor function S_n, with the sample quantile function Q_n superimposed as an inverse mapping.*

Beginning with **any** value of u in the range $\frac{j-1}{n} < u \leq \frac{j}{n}$, and then tracing $1 - u$ back to the horizontal axis, we find that $Q_n(u) = y_{(j)}$. So that we can make use of the quantile function in plots, we will conventionally select the value of u half-way between $\frac{j-1}{n}$ and $\frac{j}{n}$ to be associated with $y_{(j)}$. This conventionally selected value is at p_j, where

$$p_j = \frac{j - \frac{1}{2}}{n},$$

and is called the jth **plotting position**. Plotting positions are used in such data analytic plots as empirical survivor function plots, hazard plots, QQ-plots or probability plots. For example, following Crowder et al. (1991), the empirical survivor function is best plotted using the plotting positions:

Definition 4.2 *By an* **empirical survivor plot** *we mean a plot of plotting positions against the ranked data.* $(y_{(j)}, 1 - p_j)$, $j = 1, 2, 3, \ldots, n$. □

■ Example 4.1
Months in office for Australian Prime Ministers

In engineering applications, the lifetime (or reliability) of electronic components is of interest; in medical applications, the lifetime (or remission) of cancer patients is of interest. But what of the useful life of Prime Ministers?

Prime Minister	Term in office	Months
Barton	1 Jan 1901–24 Sept 1903	33
Deakin 1	24 Sept 1903–27 Apr 1904	8
Watson	27 Apr 1904–17 Aug 1904	5
Reid-McLean	18 Aug 1904–5 Jul 1905	12
Deakin 2	5 Jul 1905–13 Nov 1908	41
Fisher 1	13 Nov 1908–2 Jun 1909	8
Deakin 3	2 Jun 1909–29 Apr 1910	11
Fisher 2	29 Apr 1910–24 Jun 1913	39
Cook	24 Jun 1913–17 Sept 1914	16
Fisher 3	17 Sept 1914–27 Oct 1915	14
Hughes	27 Oct 1915–9 Feb 1923	89
Bruce-Page	9 Feb 1923–22 Oct 1929	81
Scullin	22 Oct 1929–6 Jan 1932	28
Lyons	6 Jan 1932–7 Apr 1939	88
Page	7 Apr 1939–26 Apr 1939	1
Menzies 1	26 Apr 1939–29 Aug 1941	29
Fadden	29 Aug 1941–7 Oct 1941	3
Curtin	7 Oct 1941–6 July 1945	46
Forde	6 July 1945–13 July 1945	1
Chifley	13 July 1945–19 Dec 1949	54
Menzies 2	19 Dec 1949–26 Jan 1966	194
Holt	26 Jan 1966–19 Dec 1967	24
McEwen	19 Dec 1967–10 Jan 1968	2
Gorton	10 Jan 1968–10 Mar 1971	39
McMahon	10 Mar 1971–5 Dec 1972	22
Whitlam	5 Dec 1972–11 Nov 1975	36
Fraser	11 Nov 1975–11 Mar 1983	89
Hawke	11 Mar 1983–19 Dec 1991	106
Keating	19 Dec 1991–2 March 1996	52

(*Source*: Cameron, R.J. (1985). *Year Book Australia 1985*, and subsequent editions.)

The data show the Australian Commonwealth Government Ministries of

SAMPLE QUANTILE FUNCTION 59

this century. Consecutive terms of office won by the same Prime Minister have been added to give total time in office.

The empirical survivor plot for these data is given in Figure 4.3. Notice that even though only one point is plotted for each ranked data point, the visual effect is quite appropriate as an estimate of a smooth survival curve.

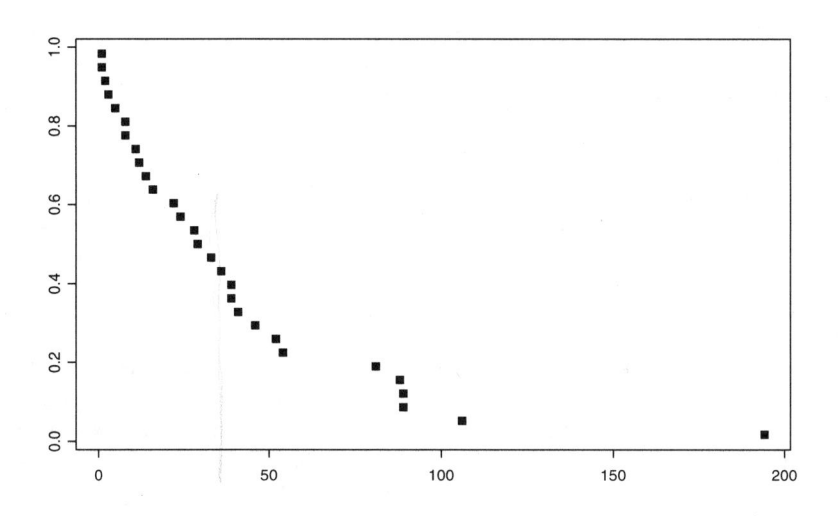

Figure 4.3 *Empirical survivor plot for the actual survival times (months in office) for Australian Prime Ministers; months on the horizontal axis and* $1 - p_j$ *(where* p_j *is the plotting position) on the vertical axis.*

□

Another choice of plotting position is

$$p_j = \frac{j}{n+1},$$

which corresponds, for $j = 1, 2, 3, \ldots, n$, to a uniformity assumption as described in the problems at the end of this chapter. The more general framework

$$p_j = \frac{i-b}{n+1-2b},$$

expressed in terms of a parameter b, is described in Blom (1958) and also in Hoaglin, Mosteller and Tukey (1983). In this text we will always take

$$p_j = \frac{j - \frac{1}{2}}{n}.$$

This means that $y_{(j)}$ is the $100(\frac{j-\frac{1}{2}}{n})$th **sample percentile** or the $\frac{j-\frac{1}{2}}{n}$th

sample quantile since

$$Q_n(p_j) = y_{(j)} \text{ for } p_j = \frac{j-\frac{1}{2}}{n}.$$

The sample quantile function estimates the population quantile function: we define this only for continuous data (for which $S = 1 - F$ is continuous and monotone decreasing):

Definition 4.3 *The pth* **population quantile**, *ξ_p is given by the solution to the equation $p = F(\xi_p)$. When ξ_p is written as a function $\xi_p = Q(p)$ of p, then Q is the* **population quantile function**. □

It is immediate from Definition 4.3 that $\xi_p = Q(p) = F^{-1}(p) = S^{-1}(1-p)$ so that Q_n is an estimator of Q.

4.3 Probability plots

A probability plot provides evidence as to whether (continuous) data have come from a population having a stated probability model S. In survival analysis, the probability models most frequently checked are the exponential and the Weibull, and these continuous distributions have been the focus of previous chapters. Since survival times are often transformed by logarithms to create a natural measurement scale against which the data may appear symmetrically distributed, normal probability plots will be relevant for the transformed data; this would mean that the original data follows a **lognormal distribution**.

For example, to assess whether the inter-arrival times of cars at a service station follow an exponential distribution, we first obtain a data set by direct observation of the service station and then decide through a stem-and-leaf plot whether an exponential distribution is a possible model.

If the probability model fits the data well, then for $p_j = \frac{j-\frac{1}{2}}{n}$, the p_jth population quantile $\xi_{p_j} = Q(p_j) = F^{-1}(p_j) = S^{-1}(1-p_j)$ of the theoretical probability distribution defined by S should be in close agreement with the p_jth sample quantile $Q_n(p_j) = y_{(j)}$. Put in other ways: based on the model S, the predicted proportion of the data that is less than the jth largest data point should be close to p_j; the predicted proportion of the data that exceeds the jth largest data point should be close to $1 - p_j$. This means that

$$F(y_{(j)}) \approx \frac{j-\frac{1}{2}}{n} \text{ and } S(y_{(j)}) \approx 1 - (\frac{j-\frac{1}{2}}{n}).$$

In order to see how close the population quantiles and the sample quantiles really are, we plot one against the other. This leads to the following definition.

Definition 4.4 *A* **probability plot, quantile-quantile plot** *or* **QQ-plot**, is a plot of the points

$$(Q(p_j), Q_n(p_j)) = \left(F^{-1}(\frac{j-\frac{1}{2}}{n}), y_{(j)}\right), \quad j = 1, 2, 3\ldots, n,$$

in the usual rectangular coordinate system. □

In a probability plot, Definition 4.4 indicates that the **ranked data** are plotted on the vertical axis and the theoretical **percentiles from a ... distribution** on the horizontal axis. The boldface terms provide appropriate labels for the axes.

If the points plotted according to Definition 4.4 fall close to a 45 degree line through the origin $(0,0)$ of the plot, then the probability model fits the data well. A **power transformation** of either (or both) axes may be required in order to produce a straight line. We will use flexibility in the interpretation of Definition 4.4. For example, a probability plot of Weibull data is most suitably presented on the logarithm scale (for each axis). The interpretation of a straight-line fit is then relative to the scale involved.

Often it is difficult to obtain perfect linear agreement at all points along the horizontal axis. Points in the plot lying above (respectively below) the 45 degree line represent data points that are larger (respectively smaller) than the probability model predicts they should be. If this occurs in the tails of the ranked data set, it may lead to an 'S-shaped plot'; that is, the bulk of the data set follows the proposed distributional form but the tails are shorter or longer than they should be.

4.4 Weibull probability plots

To check whether data have come from a $Weibull(\alpha, \beta)$ distribution, the survival function $S(y)$ of $Y \sim Weibull(\alpha, \beta)$ is required. By the arguments following (2.2),

$$S(y) = e^{-(\frac{y}{\beta})^\alpha}. \tag{4.2}$$

To construct the probability plot, substitute $y = y_{(j)}$ into each side of (4.2). Then, by taking logarithms, this reduces to

$$\log_e S(y_{(j)}) = -(\frac{y_{(j)}}{\beta})^\alpha.$$

Again, taking logarithms to create a new variable *log-life* yields

$$\log_{10}[-\log_e S(y_{(j)})] = \alpha[\log_{10} y_{(j)} - \log_{10} \beta].$$

(Note that base 10 logarithms are traditionally used to study lifetimes; however natural logarithms are equivalent, differing only by a scale change.) When log-life is made the subject of the last formula, the straight-line format

$$\log_{10} y_{(j)} = \log_{10} \beta + \frac{1}{\alpha} \log_{10}[-\log_e(S(y_{(j)}))] \tag{4.3}$$

is obtained. If the data fit the Weibull model well, then $S(y_{(j)}) \approx 1 - (\frac{j-\frac{1}{2}}{n})$. If in (4.3) we make this substitution, then when the model fits the data well,

$$\log_{10} y_{(j)} \approx \log_{10} \beta + \frac{1}{\alpha} \log_{10}[-\log_e(1 - \frac{j-\frac{1}{2}}{n})].$$

Notice that the right-side of this equation is $\log_{10} S^{-1}(1 - p_j)$. This means that for Weibull data, the plot of points

$$\left(\log_{10}[-\log_e(1 - \frac{j-\frac{1}{2}}{n})], \; \log_{10} y_{(j)}\right), \quad j = 1, 2, 3, \ldots, n, \quad (4.4)$$

would follow a straight line with slope $\frac{1}{\alpha}$ and intercept $\log_{10} \beta$; a 45 degree line results when $\alpha = 1$ and $\beta = 1$.

■ Example 4.2
Weibull age-replacements

In an article entitled *Using Statistical Thinking to Solve Maintenance Problems*, Brick et al. (1989) use a Weibull probability plot to detect whether their lifetime data on the 'time to replacement of sinker rolls in a sheet metal galvanising process' may be appropriately modelled by the Weibull distribution. This was part of a cost-cutting study. The $2000 cost of roll replacement before failure was less than half the cost of on-line failure.

$n = 17$ rollers were observed to have the following lifetimes (measured as a number of 8-hour shifts) ranked from smallest to largest in Column 3 of the table. In the final column, the log-lifetimes are given — these are plotted on the vertical axis in the probability plot against the appropriate Weibull percentiles determined in Column 1 and Column 2.

$\frac{j-\frac{1}{2}}{n}$	$\log_{10}[-\log_e(1 - \frac{j-\frac{1}{2}}{n})]$	Lifetime $y_{(j)}$	Log-lifetime $\log_{10} y_{(j)}$
0.03	−1.53	10	1.00
0.09	−1.03	12	1.08
0.15	−0.80	15	1.18
0.21	−0.64	17	1.23
0.26	−0.51	18	1.26
0.32	−0.41	18	1.26
0.38	−0.32	20	1.30
0.44	−0.24	20	1.30
0.50	−0.16	21	1.32
0.56	−0.09	21	1.32
0.62	−0.02	23	1.36
0.68	0.05	25	1.40
0.74	0.12	27	1.43
0.79	0.20	29	1.46
0.85	0.28	29	1.46
0.91	0.39	30	1.48
0.97	0.55	35	1.54

WEIBULL PROBABILITY PLOTS 63

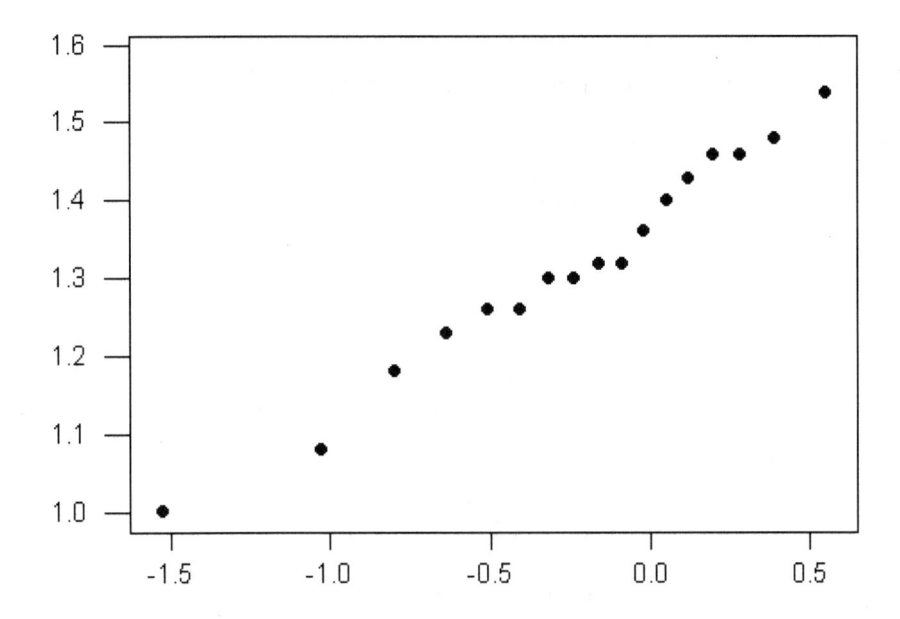

Figure 4.4 *Weibull probability plot for age-replacement; ranked log-lifetimes (base 10) on the vertical axis; percentile from Weibull distribution (log base 10 scale) on the horizontal axis.*

The plot in Figure 4.4 is reasonably assumed to be linear, although the tails of the distribution show a minor 'S' shape. To find the intercept of the straight line, set the value plotted on the horizontal axis to zero. Solving

$$\log_{10}[-\log_e(1 - \frac{j - \frac{1}{2}}{n})] = 0,$$

it follows that

$$\frac{j - \frac{1}{2}}{17} = 1 - \frac{1}{e} = 0.63 \text{ (2 dp)}.$$

The 0.63 quantile, or 63rd sample percentile, is approximately 23 and this provides a reasonable estimate of β, the scale parameter. For estimating α, we take the inverse of the slope of the estimated straight line: this gives an α value of approximately 3.33. Since Weibull models with $\alpha > 1$ have hazard functions that increase with time, and because of the costs involved, the authors of the article indicate that it is inappropriate to wait for the sinker rolls to fail. A policy of scheduled replacement should be used. □

The slope and intercept of a well-fitting line (perhaps obtained by standard linear regression techniques such as the method of least squares) may provide estimates for $\frac{1}{\alpha}$ and $\log_{10} \beta$ (and hence for α and β.) Notice that if the observed slope of a fitted line is close to 1, then the estimate of α is close to 1, and

4.5 Normal probability plots for the lognormal

In this section we concentrate on checking whether data may be assumed to be lognormally distributed. According to Definition 2.6, the survival function for a $lognormal(\mu, \sigma)$ distribution is given by

$$S(y) = 1 - \Phi(\frac{\log_e y - \mu}{\sigma}),$$

where

$$\Phi(x) = \int_{-\infty}^{x} \frac{1}{\sqrt{2\pi}} e^{-\frac{1}{2}t^2} dt.$$

Placing $y = y_{(j)}$ in each side of this equation we obtain

$$F(y_{(j)}) = \Phi(\frac{\log_e y_{(j)} - \mu}{\sigma}).$$

Therefore, if F is the cdf for a probability model that fits the data well, then $F(y_{(j)}) \approx \frac{j-\frac{1}{2}}{n}$. This identification on the left-side of our equation gives

$$\frac{j-\frac{1}{2}}{n} = \Phi(\frac{\log_e y_{(j)} - \mu}{\sigma}).$$

On taking Φ^{-1} of each side of this equation,

$$\Phi^{-1}(\frac{j-\frac{1}{2}}{n}) = \frac{\log_e y_{(j)} - \mu}{\sigma},$$

or

$$\log_e y_{(j)} = \mu + \sigma \Phi^{-1}(\frac{j-\frac{1}{2}}{n}).$$

This means that, for lognormally distributed data, the plot of points

$$\left(\Phi^{-1}(\frac{j-\frac{1}{2}}{n}), \log_e y_{(j)} \right), \quad j = 1, 2, 3, \ldots, n,$$

would follow a straight line with intercept μ and slope σ. As in the Weibull plot, such parameters may then be *estimated* from the slope and intercept of the observed plotted line when it is agreed that the data are from a lognormal population. This estimation process is best seen as a guide to the magnitudes of mean and variance rather than an accurate estimation procedure.

The inverse cdf of the normal distribution may need to be hand-calculated: for example, tables of the standard normal distribution show that for a sample of size $n = 5$,

HAZARD PLOTS

i	$\frac{i-\frac{1}{2}}{n}$	$\Phi^{-1}(\frac{i-\frac{1}{2}}{n})$
1	0.1	-1.28
2	0.3	-0.52
3	0.5	0.00
4	0.7	0.52
5	0.9	1.28

Notice that when data have been grouped into intervals in a frequency distribution so that the identity of the original data is lost, the plot may still proceed by using the interval endpoints and the sample percentiles that these endpoints correspond to. Thus, if interval endpoint e_i corresponds to the $100(\frac{j-\frac{1}{2}}{n})$th sample percentile (where j depends on the value of i), the points

$$\left(F^{-1}(\frac{j-\frac{1}{2}}{n}), \ e_{(i)} \right), \quad i = 1, 2, 3 \ldots, k$$

are plotted; here, k denotes the number of interval endpoints.

4.6 Hazard plots

The following theorem, commonly called the **delta method**, is used to determine the approximate mean and variance of a transformation $g(Y)$ of a survival random variable Y when the mean and variance of Y are known. The theorem may be usefully applied in many different contexts. In this instance, it will be used to determine the mean and variance of a simple estimate of the cumulative hazard function.

Theorem 4.1 *(The Delta Method)* *If Y is a survival random variable with $E(Y) = \mu$ and $Var(Y) = \sigma^2$, then for a function g with $g'(\mu) \neq 0$, the transformed random variable $g(Y)$ has approximate mean and variance given by $E[g(Y)] \approx g(\mu)$ and $Var[g(Y)] \approx \sigma^2 [g'(\mu)]^2$.* □

An intuitive understanding of the proof of this theorem is as follows: following Miller (1981), use Taylor's Theorem to expand $g(Y)$ about μ.

$$g(Y) = g(\mu) + (Y - \mu) \frac{g'(\mu)}{1!} + (Y - \mu)^2 \frac{g''(\mu)}{2!} + \ldots .$$

If we use only a linear approximation, then $g(Y) \approx g(\mu) + (Y - \mu) g'(\mu)$, and the result follows by taking the mean and variance of each side of this equation (making sure that $\mu, g(\mu)$ and $g'(\mu)$ are treated as constants).

This theorem will be easy to apply to the cumulative hazard function, which is a simple power transformation of the survival function.

Definition 4.5 *The **cumulative hazard function**, H, for a survival random variable with survival function $S(y)$ is given by $H(y) = \log_e S(y)$.* □

According to Theorem 1.3, H may be expressed as

$$H(y) = \int_0^y h(u)du$$

and this definition shows the concept of the hazard being accumulated through integration.

Since S_n is an appropriate estimator of S, it follows that

$$H_n(y) = -\log_e S_n(y)$$

is an obvious estimator of

$$H(y) = -\log_e S(y).$$

H_n is called the **empirical cumulative hazard function** for estimating H. The accuracy of such an estimate may be assessed through the Delta Method of Theorem 4.1.

To see this we need to make the following identification in Theorem 4.1: put $Y = S_n(y)$; then

$$g(Y) = H_n(y) = -\log_e S_n(y) = -\log_e Y$$

is the choice of function g satisfying $g'(Y) = -\frac{1}{Y}$. Therefore,

$$E[g(Y)] = E[H_n(y)] \approx -\log_e E[Y] = -\log_e S(y) = H(y)$$

and

$$Var[g(Y)] = Var[H_n(y)] \approx \left(\frac{1}{E(Y)}\right)^2 Var(Y) = \frac{1}{S(y)^2}\frac{S(y)[1-S(y)]}{n}.$$

Taking square roots gives a standard error which is estimated by replacing S by S_n. Then, at a fixed point $y = y^*$, the cumulative hazard is estimated by the approximate two standard error confidence interval

$$H_n(y^*) \pm 2\sqrt{\frac{1-S_n(y^*)}{nS_n(y^*)}}.$$

If the cumulative hazard function is evaluated at the observed ranked data, the expected result is not only surprising, but also very useful as a simple alternative to probability plots.

Theorem 4.2 *The cumulative hazard function of a survival random variable satisfies* $(j = 1, 2, 3, \ldots, n)$

$$E\left\{H[Y_{(j)}]\right\} = \sum_{i=1}^{j} \frac{1}{n-i+1}.$$

□

The proof of this result is left for the exercises. (It is worth noting that it depends on the result in Theorem 3.7.) The quantities on the right-side of the equation in Theorem 4.2 have a special role in hazard plots.

EXERCISES

Definition 4.6 *A* **hazard plot** *is a simple plot of the points* $(\alpha_j, y_{(j)})$, *where*

$$\alpha_j = \sum_{i=1}^{j} \frac{1}{n-i+1}$$

are called the **hazard plot scores**. □

To use a hazard plot to detect if data are exponential, note that

$$h(y) = \frac{1}{\beta}.$$

It follows that $H(y) = \int_0^y h(u)du = \int_0^u \frac{1}{\beta}du = \frac{y}{\beta}$. Therefore, if the data have come from an exponential distribution, the relationship between α_j and $y_{(j)}$ should be a straight line. Many engineers regard this as a simpler diagnostic test than a probability plot. Similarly, if the data have come from a Weibull distribution, the relationship between $\log_e \alpha_j$ and $\log_e y_{(j)}$ should be a straight line.

4.7 Exercises

4.1. Estimate the probability of survival beyond a full 3-year term in office for the 'Months in office for Australian Prime Ministers' data of Example 4.1. Place appropriate error bounds on your estimate.

4.2. Construct the empirical distribution function for 'Survival (days)' for the patients observed to die from 'heart failure' in the 1980 Stanford Heart Transplant Data from Crowley and Hu (1977). (The complete data set will be discussed in later chapters and is listed in Example 5.2.)

Give a point and interval estimate for the probability of survival beyond 1000 days.

4.3. Construct a hazard plot for 'Survival (days)' for the patients observed to die from 'heart failure' in the 1980 Stanford Heart Transplant Data from Crowley and Hu (1977). (The complete data set will be discussed in later chapters and is listed in Example 5.2.)

Give an approximate point and interval estimate of the cumulative hazard at 1000 days.

Use a probability plot to assess whether survival time after transplant follows a lognormal distribution.

4.4. There is not complete agreement between authors on the 'optimal' representation of sample percentiles. One family of representations may be described as follows: $x_{(i)}$ is the $100(\frac{i-b}{n+1-2b})$th sample percentile. Popular choices of b are $b = 0, \frac{1}{2}, \frac{3}{8}$. For large data sets there appears to be little difference between the choices for b.

Which value of b corresponds to the definition of plotting position used throughout this book?

4.5. Lawless (1982), Gehan (1965) and others have discussed data from a clinical trial examining steroid-induced remission times (weeks) for leukemia patients. One group of 21 patients were given 6-mercaptopurine (6-MP); a second group of 21 patients were given a placebo. Since the trial lasted 1 year and patients were admitted to the trial during the year, some of the data could not be gathered by the cut-off date when some patients were still in remission. (This lack of completion of the data is termed 'right-censoring' and is fully discussed in the next chapter.) Observations on remission time Y were:

6-MP		Placebo	
6	17+	1	8
6	19+	1	8
6	20+	2	11
6+	22	2	11
7	23	3	12
9+	25+	4	12
10	32+	4	15
10+	32+	5	17
11+	34+	5	22
13	35+	8	23
16		8	

(*Source*: Lawless, J.F. (1982). *Statistical Models and Methods for Lifetime Data*. Wiley: New York.)

Notation: '+' denotes 'right censoring' in the 6-MP group, so that 6+ represents an observed 6-week remission which was still in effect at the closure of the trial. Censoring is discussed in the next chapter. For now we will concentrate on the placebo data.

Use these data to construct a graph of the empirical survivor function for the placebo group for which all the data are completely observed.

Now construct an empirical survivor plot for the data — note the shape of the implied survival curve.

4.6. Determine the population quantiles $Q(p_j), j = 1, 2, 3, \ldots, 10$, for an exponential probability model with mean 1.

4.7. Determine the sample quantiles $Q_{10}(p_j), j = 1, 2, 3, \ldots, 10$, in terms of the members of an ordered sample $y_{(1)}, y_{(2)}, \ldots, y_{(10)}$ of possibly exponentially distributed data.

EXERCISES

4.8. Show that $\xi_p = F^{-1}(p) = S^{-1}(1-p)$.

4.9. Determine formulae for the plotted points in a QQ-plot used to check whether data follow an $exp(\beta)$ model. For exponentially distributed data, how can the QQ-plot be used to estimate β?

4.10. What shape would you expect for the cumulative hazard $H(y)$ representing a survival variable which is exponentially distributed? What shape results if the survival variable follows a Weibull distribution?

Discuss the effect of these results in plots of the empirical cumulative hazard function.

4.11. The lifetime of an article is thought to have an exponential distribution. Twelve such articles were selected at random and tested until nine of them failed. The nine observed failure times were:

$$8, 14, 23, 32, 46, 57, 69, 88 \text{ and } 109.$$

Check the conjecture of exponentially distributed lifetimes.

4.12. The listed data give failure times of an oil-fired boiler formerly located at the Brighton 'B' power station in southeastern England. The data are rounded to the nearest day and the high frequency at 1 is caused by failures in the first day.

Failure times (days)

Time	Freq.	Time	Freq.	Time	Freq.	Time	Freq.
1	25	14	2	34	2	58	1
2	12	15	2	35	1	61	2
3	5	18	2	36	2	64	1
5	2	19	1	40	1	68	1
6	3	22	1	41	1	77	1
7	3	23	3	44	1	81	1
8	2	27	2	49	1	86	1
9	4	28	3	50	1	87	3
10	3	29	1	54	1	95	1
11	1	31	2	55	1		
12	2	32	1	56	1		

(*Source*: Baxter, L.A. and Li, L. (1996). Nonparametric estimation of limiting availability. *Lifetime Data Analysis*, 2, 391–403.)

What probability model seems most appropriate for these data? Discuss whether your chosen model appears to fit the data well, for example, in the centre and in the tails of the distribution. Do there appear to be any outliers or atypical values in the data?

4.13. The cumulative distribution function of a $Weibull(\alpha, \beta)$ distribution is given by
$$F(y) = 1 - e^{-(\frac{y}{\beta})^\alpha}.$$

(a) Substitute $y = F^{-1}(x)$ into each side of this equation and hence obtain a formula for $F^{-1}(x)$ in terms of x.

(b) Show that $\log_{10} F^{-1}(x)$ is a linear function of $\log_{10}[-\log_e(1-x)]$.

(c) Discuss these results in the context of a Weibull probability plot.

4.14. The following data (Chatfield, 1983) resulted from a life test of refrigerator motors (hours to burnout):

Hours to burnout				
104.3	158.7	193.7	201.3	206.2
227.8	249.1	307.8	311.5	329.6
358.5	364.3	370.4	380.5	394.6
426.2	434.1	552.6	594.0	691.5

(Source: Chatfield, C. (1983). *Statistics for Technology — A Course in Applied Statistics*, 3rd ed. Chapman & Hall: London.)

(a) Construct a Weibull probability plot for these data.

(b) Estimate the parameters α and β in the fitted probability model.

(c) Use the fitted model to estimate the probability that a refrigerator motor will fail in less than 300 hours.

(d) If there is a threshold lifetime t_0 beyond which all motors will last, how could t_0 be estimated from the data? Subtract this estimated value of t_0 from all the data and reconstruct a Weibull probability plot for all non-zero data points.

(e) Estimate the values of t_0, α_0 and β_0 in the model
$$P(T > t) = e^{-(\frac{t-t_0}{\beta_0})^{\alpha_0}},$$
specifying the probability that the lifetime, T, of a refrigerator motor exceeds the time t.

4.15. Suppose that a large number of items are put on an industrial life test. The time until failure of these items is not recorded directly, but rather, as time passes, at the end of each time unit (such as 1 day, 1 week, ...) the number of surviving items is recorded. Suppose that at the end of time unit i, there are n_i items remaining.

The key feature here is that there are a large number of items and we do not observe the failure time of each item.

If we suspect that the lifetimes follow a Weibull distribution with survival function
$$S(y) = e^{-(\frac{y}{\beta})^\alpha},$$
then how can we estimate the parameters α and β without observing directly any of the data? We assume that observing continues until all items have failed.

(a) Let the time periods be denoted by $i = 1, 2, 3, \ldots, n$. Assuming a Weibull model, show that if we estimate $S(y)$ at time period i, or $y = i$, by $\frac{n_i}{n}$, then a plot of points
$$\left(\log_e[-\log_e(\frac{n_i}{n})], \ \log_e i\right), \quad i = 1, 2, 3, \ldots, n,$$
would have least squares slope an estimate of $\frac{1}{\alpha}$ and intercept an estimate of $\log_e \beta$.

(b) How does the plot of points in (a) compare with the Weibull probability plot of equation (4.4)?

(c) Mendenhall and Sincich (1992) indicate that according to a 'weakest link' hypothesis, a piece of industrial equipment may fail when the first of multiple components in that equipment fails and that a Weibull probability model for the equipment's lifetime is often found appropriate in these circumstances. The following data relate to the failure times of $n = 138$ roller bearings.

Hours (hundreds) i	1	2	3	4	5	6	7
Surviving bearings n_i	138	114	104	64	37	29	20

Hours (hundreds) i	8	12	13	17	19	24	51
Surviving bearings n_i	10	8	6	4	3	2	1

(Source: Nelson, W. (1972). Weibull analysis of reliability data with few or no failures. *Journal of Quality Technology*, 17, 141-2.)

Use the plotting procedure of (a) to construct estimates for α and β in the Weibull model.

4.16. Suppose that Y is a survival random variable with survival function S. Show that $S(Y)$ is uniformly distributed on the interval $[0, 1]$. In this notation, $S(Y)$ is a function of the random variable Y and is therefore a random variable. (Hint: Show that the random variable $W = S(Y)$ has cumulative distribution function $F_W(w) = w$ for $0 \leq w \leq 1$.)

4.17. Suppose that Y is a survival random variable with survival function S and cumulative hazard function $H(y) = -\log_e S(y)$. Show that
$$H(Y) \sim exp(1).$$

In this notation, $H(Y)$ is a function of the random variable Y and is therefore a random variable. (*Hint*: This result builds on the ideas of the previous problem.)

4.18. Let $Y_1, Y_2, \ldots, Y_n$ be independent and identically distributed random variables with survival function S and cumulative hazard function H. Show that $H[Y_{(i)}], i = 1, 2, 3, \ldots, n,$ is an ordered sample from an exponential population with mean 1. (*Hint*: This result builds on the ideas of the previous problem.)

4.19. Let $Y_1, Y_2, \ldots, Y_n$ be independent and identically distributed random variables with survival function S and cumulative hazard function H. Show that

$$E\left[H[Y_{(i)}]\right] = \sum_{j=1}^{i} \frac{1}{n-j+1}.$$

(*Hint*: This result builds on the ideas of the previous problem.)

4.20. Show that the ith largest of n observations randomly chosen from the unit interval has mean $\frac{i}{n+1}$.

Suppose that observations $Y_1, Y_2, \ldots, Y_n$ are independent and identically distributed with cumulative distribution function F. Show that

$$E[F(Y_{(i)})] = \frac{i}{n+1}.$$

This is the justification of the use by some authors of $\frac{i}{n+1}$ as a choice of plotting position.

CHAPTER 5

Censoring and Lifetables

Censoring occurs when we are unable to observe the response variable of interest. For example, in a clinical trial monitoring the remission of cancer patients, some patients may still be in remission at the trial's closure, so that for these patients, we know only that their true period of remission was longer than the duration of the trial. Such an observed time is a **censored lifetime**. In another example, the patient may withdraw from the trial and move out of town (or overseas) — such patients are **withdrawals** and their status is described as **lost to follow-up**, meaning that their progress can no longer be followed; again, their true remission is longer than could be observed. Such patients are **right-censored**, since on a timeline their true lifetimes are to the right of their observed censor times.

On the other hand, Finkelstein (1986) describes sacrifice experiments in experimental animals (such as mice). Of interest is the time from injection of a carcinogen until the onset of tumors. An attempt to measure this time is made by sacrificing the animal, and observing at this time of death if tumors are present — in which case the required onset time is smaller than the censored time actually observed and the observation is **left-censored** — if tumors are absent, time to tumor onset is right-censored. This shows how both left- and right-censored data may occur in the same data set.

These are examples of what are called Type I censored data. For such censored data, the actual number of censored data points is unknown in advance, and so is a random variable. For example, we do not know at the outset of the experiment how many of Finkelstein's mice will show tumors at time of sacrifice. The data of Example 5.1 are extracted from a newspaper report showing a very graphic example of Type I censoring from the media (as distinct from censoring of the media).

■ **Example 5.1**
Western hostages held for more than 6 months since 1984
The Australian, Wednesday Nov 20, 1991, reported data on 'Western hostages held for more than 6 months since 1984.' Post-1984, and for a period lasting into the early 1990s, more than 30 Westerners were kidnapped by various fundamentalist groups in the Middle East; each group apparently had its own agenda, but many operated under the umbrella of the political/religious group *Hezbollah*.

Some of the aspects of these data are reproduced in the diagram in Figure 5.1. Individual case histories are tracked in timelines. Of interest is the variable 'Time to release'. Right-censoring occurred when some hostages had not been released by the time of the news report, or had died from other 'causes.' □

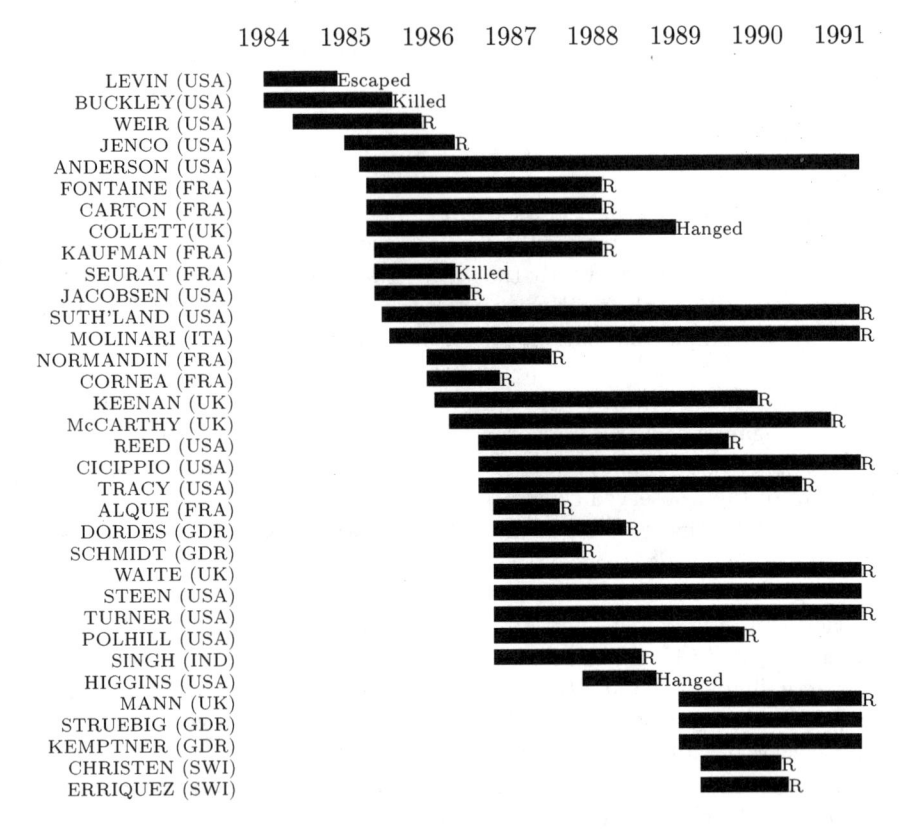

Figure 5.1 *Timelines from initial capture until release, denoted by 'R', or censor time caused by death or lack of release by 1991 for 34 Western hostages.* Source: The Australian, *Wednesday November 20, 1991.*

5.1 Truncation

Type I censoring is characterised by the experimentalist's lack of control over how much data are observed in censored form; we cannot prevent people leaving a clinical trial should they wish to do so. However, in all cases we can identify which data are observed in censored form, and which data are observed as true realised lifetimes. This is very different from the case of **truncated data**, where a truncated data point simply cannot be seen so that we may not even know of its existence. The difference seen by contrasting examples: the survival time of a heart transplant patient is right-censored if the patient is killed in a car accident but the patient has been monitored up

TYPE I CENSORING

to the time of the crash; with a given telescope, we can only detect a very distant stellar object which is brighter than some limiting flux — the object is left-truncated if it lies beyond detection by our telescope — we cannot tell if the object is even there if we cannot see it. (For further discussion of censoring and truncation in astronomical data, see Feigelson and Babu, 1996.)

Definition 5.1 *From a population of an unknown number N possible observations, we say that the data set $Y_1, Y_2, Y_3 \ldots, Y_n$ of $n < N$ elements is* **left-truncated** *by the fixed truncation constants $t_1, t_2, t_3, \ldots, t_n$ if the observed sample consists of ordered pairs (y_i, t_i) where $y_i \geq t_i$, so that y_i is left-truncated by t_i for each $i = 1, 2, \ldots, n$.* □

Assumption: $Y_1, Y_2, Y_3, \ldots, Y_n$ are assumed to be independent of the mechanism generating the $t_1, t_2, t_3, \ldots, t_n$.

In the astronomical example discussed earlier, $t_1 = t_2 = t_3 = \ldots = t_n = t$, say, the threshold magnitude limit of light from a distant object that the given telescope may detect. **Right-truncation** is defined similarly in that, with the same notation as Definition 5.1, the observed sample consists of ordered pairs (y_i, t_i) where $y_i \leq t_i$.

5.2 Type I censoring

We return now to the classification of Type I and Type II censoring. In Type I censoring, the number of uncensored observations is a random variable, while in Type II censoring, the number of uncensored observations is fixed in advance. These concepts are made clear from the notation and definitions which follow. Suppose that $Y_1, Y_2, Y_3, \ldots, Y_n$ are independent and identically distributed survival variables.

Definition 5.2 *The survival variables $Y_1, Y_2, Y_3, \ldots, Y_n$ are* **right-censored** *by fixed constants $t_1, t_2, t_3, \ldots, t_n$, if the observed sample consists of the ordered pairs (Z_i, δ_i), for $i = 1, 2, \ldots, n$, where for each i: $Z_i = \min\{Y_i, t_i\}$,*

$$\delta_i = \begin{cases} 1 & \text{if } Y_i \leq t_i \quad \text{(uncensored)} \\ 0 & \text{if } Y_i > t_i \quad \text{(censored)}, \end{cases}$$

where t_i is the fixed **censor time** *and δ_i the* **censor indicator** *for Y_i.* □

Assumption: $Y_1, Y_2, Y_3, \ldots, Y_n$ are assumed to be independent of the mechanism generating the $t_1, t_2, t_3, \ldots, t_n$.

Note that according to this definition, t_i is observed when Y_i is censored. If Y_i is uncensored, and corresponds to a true observed death time, then t_i often remains unknown.

Example 5.2
Stanford Heart Transplant Data
This example gives data from the Stanford Heart Transplant program.

No	Days	Cens	Age	T5	No	Days	Cens	Age	T5
1	15	1	54.3	1.11	35	322	1	48.1	1.82
2	3	1	40.4	1.66	36	838	0	41.6	0.19
3	624	1	51.0	1.32	37	65	1	49.1	0.66
4	46	1	42.5	0.61	38	815	0	32.7	1.93
5	127	1	48.0	0.36	39	551	1	48.9	0.12
6	64	1	54.6	1.89	40	66	1	51.3	1.12
7	1350	1	54.1	0.87	41	228	1	19.7	1.02
8	280	1	49.5	1.12	42	65	1	45.2	1.68
9	23	1	56.9	2.05	43	660	0	48.0	1.20
10	10	1	55.3	2.76	44	25	1	53.0	1.68
11	1024	1	43.4	1.13	45	589	0	47.5	0.97
12	39	1	42.8	1.38	46	592	0	26.7	1.46
13	730	1	58.4	0.96	47	63	1	56.4	2.16
14	136	1	52.0	1.62	48	12	1	29.2	0.61
15	1775	0	33.3	1.06	49	499	0	52.2	1.70
16	1	1	54.2	0.47	50	305	0	49.3	0.81
17	836	1	45.0	1.58	51	29	1	54.0	1.08
18	60	1	64.5	0.69	52	456	0	46.5	1.41
19	1536	0	49.0	0.91	53	439	0	52.9	1.94
20	1549	0	40.6	0.38	54	48	1	53.4	3.05
21	54	1	49.0	2.09	55	297	1	42.8	0.60
22	47	1	61.5	0.87	56	389	0	48.9	1.44
23	51	1	50.5	9999	57	50	1	46.4	2.25
24	1367	0	48.6	0.75	58	339	0	54.4	0.68
25	1264	0	45.5	0.98	59	68	1	51.4	1.33
26	44	1	36.2	0.00	60	26	1	52.5	0.82
27	994	1	48.6	0.81	61	30	0	45.8	0.16
28	51	1	47.2	1.38	62	237	0	47.8	0.33
29	1106	0	36.8	1.35	63	161	1	43.8	1.20
30	897	1	46.1	9999	64	14	1	40.3	9999
31	253	1	48.8	1.08	65	167	0	26.7	0.46
32	147	1	47.5	9999	66	110	0	23.7	1.78
33	51	1	52.5	1.51	67	13	0	28.9	0.77
34	875	0	38.9	0.98	68	1	0	35.2	0.67
					69	1	1	41.5	0.87

(*Source*: Crowley, J. and Hu, M. (1977). The covariance analysis of heart transplant data. *Journal of the American Statistical Association*, 72, 27–36.)

This program began in the pioneering days of heart transplants on 1 October, 1967 and by the initial cut-off date 1 April 1974, 69 patients had received heart transplants. By February 1980, some 184 patients had received transplants (Crowley and Hu, 1977). Tracking the success of these transplants has now become part of medical history.

In these data, 24 patients were alive at the cut-off date and constitute censored observations. This was caused in part by doctors admitting younger patients into the program for ethical reasons: the younger patients tended to show better survival prospects. However, the independence assumption relating lifetimes and censoring mechanism was violated in the process.

Other explanatory information provided in these data are the **mismatch score**, **T5**, and **Age**. The mismatch score is a numerical measure of closeness of alignment between the donor and recipient tissue. The effect of the mismatch score on survival is examined using the regression techniques of later chapters.

In the notation of Definition 5.2, the observed censored heart transplant data are the 69 ordered pairs: $(15, 1), (3, 1), (624, 1), (46, 1), \ldots$.

▰ Example 5.3
Patient monitoring after Astrocytomas diagnosis
The accompanying data from the Anti-Cancer Council of Victoria show survival time in months from date of diagnosis of Astrocytomas (brain tumors) until death resulting from the effects of tumors.

Astrocytomas Survival Time (Months)

7	9+	58+	60+	10+	0	23+	1+	54	40
124	60	6+	2+	11	13	3	1	1+	70
88+	65	84+	83+	73+	2	4	11	11	34+
13	2+	2	5+	5	61	2	96+	1	53+
2+	14	13	5+	46+	46+	2+	30+	13+	18+
79	8	43	40+	11	9	2	1+	4+	15
17	2+	4	1+	3+	26+	12	13	5	105+
4+	2+	19	1	2	24+	60	4	65+	67+
5	1+	68+	15	1+	24+	2	1	83+	101+
10	9	39+	13+	35+	28+	27+	1	63	44+
21	35	19+	20+	20+	1	8	6+	14	12
53+	63+	15+	40+	14	53	29+	119+	76	27+
2	26+	26+	8	24+	25+	26+	19+	18+	14+
41+	3	4	13	6	3	16	30	13	33+

(*Source*: M. Staples, The Anti-Cancer Council of Victoria, Australia. *Private communication*, 1994.)

The + notation represents right-censoring caused by withdrawal from the ongoing medical monitoring process, or a recording of the current status as 'alive'. Almost half the subjects were in this last category as patients were admitted into the study at different stages depending on the date of original diagnosis. These data were part of a larger ongoing study of brain tumor patients in Victoria and New South Wales.

For **left-censored data**, the observed lifetimes are $Z_i = \max\{Y_i, l_i\}$, where l_i is the (left-) censor time associated with Y_i. Since, for left-censored data,

$$-Z_i = \min\{-Y_i, -l_i\},$$

it follows that left-censoring is a special case of right-censoring with the time axis reversed. It is because of this that there have been few specialist techniques developed explicitly for left-censored data.

Double-censoring occurs (see, for example, Turbull, 1974) as a combination of left- and right-censoring. Here we observe

$$Z_i = \max\{\min\{Y_i, t_i\}, l_i\},$$

where l_i and t_i are the left-, respectively right-, censor times associated with Y_i for $l_i < t_i$. In this case Y_i is only observed if it falls in a window of observation $[l_i, t_i]$, otherwise one of the endpoints of the window is observed and the other endpoint of the window (probably) remains undisclosed.

This type of censoring is not to be confused with more general **interval-censoring**, where the observed data consists of intervals $I_1, I_2, \ldots, I_n$, where, for each $i = 1, 2, \ldots, n$, the ith response lies somewhere in interval I_i. (See, for example, Smith, 1996.) In this case an uncensored observation of an observed death corresponds to an observed 'interval' consisting of a single point. For example, in a clinical trial where the effect of a new drug on patient remission times is being assessed, if the ith patient is in remission at 6 weeks into the trial, but misses future weekly checkups until the ninth week, when the patient is found to be out of remission, then $I_i = [6, 9)$ is the observation of the patient's 'length of remission'.

Cohen (1991) gives an impressive list of parametric procedures to use when analysing many types of censored and truncated data.

■ Example 5.4
Median descent time of a baboon troop

The following times were recorded in a famous study by Wagner and Altman (1973), where their paper posed the question: What time do the baboons come down from the trees?

A troop of baboons in the Amboseli Reserve in Kenya were found to sleep in trees each night and during the morning descend from the trees to begin the day's work of foraging for food. Observers arrived each day to detect the time by which half the troop had descended (**median descent time**). On some days (Table A) they observed this process and recorded the desired median descent time. However, on the days recorded in Table B, they arrived too late to observe this time as more than half the troop had already descended.

TYPE II CENSORING

Table A: Median descent time

0656	0754	0820	0833	0848	0904	0920	0952
0659	0758	0820	0836	0850	0905	0930	1027
0720	0805	0825	0840	0855	0905	0930	
0721	0808	0827	0842	0858	0907	0932	
0743	0810	0828	0844	0858	0908	0935	
0747	0811	0831	0844	0859	0910	0935	
0750	0815	0832	0845	0859	0910	0945	
0751	0815	0832	0846	0900	0915	0948	

Table B: Censored arrival time

0705	0800	0815	0855	0915	0955	1245	1653
0710	0805	0817	0856	0920	1005	1250	1705
0715	0805	0823	0857	0920	1012	1405	1708
0720	0805	0830	0858	0925	1018	1407	1722
0720	0807	0831	0858	0926	1020	1500	1728
0730	0810	0838	0858	0931	1020	1531	1730
0740	0812	0840	0859	0933	1020	1535	1730
0750	0812	0840	0900	0943	1031	1556	1750
0750	0813	0845	0905	0945	1050	1603	1801
0753	0814	0850	0907	0946	1050	1605	1829
0755	0815	0851	0908	0950	1100	1625	
0757	0815	0853	0915	0955	1205	1625	

(*Source*: Wagner, S. S. and Altman, S. A. (1973). What time do the baboons come down from the trees? (An estimation problem). *Biometrics*, 29, 623-35.)

The times are recorded on a 24-hour clock. The information in Table A represents observed times to the required event 'half the troop has descended from the trees'. The information in Table B represents left-censored observations of the required time to event since the actual times were smaller than those recorded. □

5.3 Type II censoring

Suppose that, for a given sample $Y_1, Y_2, Y_3, \ldots, Y_n$ of size n, only the first $r < n$ lifetimes are observed. The value of r is fixed before the survival data are seen. This means that the observed data consist of the *smallest* r observations. In terms of random variables, this may be expressed using **order statistics**: from the possible responses $Y_1, Y_2, Y_3, \ldots, Y_n$ we observe only the first r ranked responses $Y_{(1)}, Y_{(2)}, \ldots, Y_{(r)}$. Recall from Chapter 4 that this notation means that

$$y_{(1)} \leq y_{(2)} \leq y_{(3)} \leq \cdots \leq y_{(r-1)} \leq y_{(r)}$$

as realised data. This is the case which occurs in Example 5.5, where because of extreme time and cost considerations, a **lifetest** of aircraft components cannot necessarily wait until all components have failed; it may take longer than experimental circumstances permit.

Example 5.5
Lifetesting aircraft components

Mann and Fertig (1973) provide data on the failure times of aircraft components subject to a lifetest. The data are obtained from $n = 13$ randomly selected test items and the lifetest terminated at the observed failure of the 10th item. The $r = 10$ observed lifetimes were (in hours)

$$0.22,\ 0.50,\ 0.88,\ 1.00,\ 1.32,\ 1.33,\ 1.54,\ 1.76,\ 2.50,\ 3.00.$$

These data have been discussed by many authors, including Cohen (1991) and Lawless (1982), who fitted a Weibull model to these data. □

Type II censoring is a form of right-censoring because we know of the censored lifetimes only that they exceed $y_{(r)}$. This type of censoring is commonly encountered in reliability studies in engineering. Note that the duration of such a lifetest is not fixed in advance, because before the experiment commences we cannot tell how long it will take for the rth test item to fail.

Lawless (1982) notes that Type II censored data are easily plotted in a probability plot. If it is conjectured that the data $y_1, y_2, y_3, \ldots, y_n$ are from a probability model with survival function S, then following the theory of probability plots in Chapter 4, the probability plot for Type II censored data consists of the r points

$$\left(S^{-1}(1 - \frac{j - \frac{1}{2}}{n}), y_{(j)}\right),\ j = 1, 2, \ldots, r,$$

based on all n plotting positions $p_j = \frac{j-\frac{1}{2}}{n}$. Therefore, when making a probability plot for Type II censored data, although plotting positions are obtained for all n points, only the first r observations are actually plotted. This all works appropriately because a probability plot is based on ranked data.

5.4 Lifetable estimates

Suppose that $Y_1, Y_2, Y_3, \ldots, Y_n$ are independent and identically distributed random variables with survival function S and probability density function f.

We present here a method, called the **cohort lifetable method**, for estimating the probability of survival beyond a fixed point in the presence of censored data. This is one of the oldest and widely used methods for the nonparametric analysis and presentation of right-censored data. The lifetable method has grown traditionally from the medical and actuarial contexts. It is especially useful for estimation with large amounts of data. When there is no censoring or even Type II censoring, the empirical survivor function is easily plotted to provide information on the lifetime distribution and the shape of

LIFETABLE ESTIMATES

S. To cater for Type I censoring, we follow the approach of Lawless (1982) to show how a lifetable analysis may be achieved in three simple steps. We also demonstrate how a lifetable is realised using statistical software. First, we develop the lifetable structure and notation.

Three steps to a lifetable analysis:

1. Divide the lifetime axis into fixed disjoint intervals. (don't have to be same width)
2. Estimate the conditional probability of survival across each interval.
3. Estimate S at the interval endpoints.

We now expand on how each of these three stages is actually achieved. For Step 1: Begin by imagining n patients alive and under observation as time passes and records are kept of their survival time (time of death) or censor time. To keep things simple, these records are kept in interval format. We begin our construction by dividing $[0, \infty)$, the support of the lifetime distribution, into fixed disjoint intervals

$$I_j = [a_{j-1}, a_j), \quad j = 1, 2, 3, \ldots, k+1 \quad \text{such that} \quad a_0 = 0 \quad \text{and} \quad a_{k+1} = \infty.$$

The a_j's are interval endpoints chosen by the data analyst in much the same way as histogram intervals are chosen — a choice of too few results in limited accuracy. Having constructed the intervals, we monitor the data (censor times and death times) occurring in the intervals. One of the most important lifetable concepts is that of **number at risk** in a given interval.

Definition 5.3 *In a lifetable, the **number at risk** in any interval is the number alive and under observation and not censored at the start of that interval.* □

In terms of notation for the jth interval $I_j = [a_{j-1}, a_j)$, we write:

N_j as the **number at risk** in I_j;
D_j as the **number of deaths**, or observed failures in I_j;
W_j as the **number censored** in I_j.

Those censored in I_j are also termed **withdrawals**, or **losses to follow-up** in I_j. These definitions imply that $N_1 = n$, so that the entire sample is initially at risk. Clearly $N_j = N_{j-1} - D_{j-1} - W_{j-1}$ demonstrates how those at risk in the $(j-1)$th interval propagate into the jth interval. This propogation is at the heart of a lifetable analysis, making the lifetable a chart showing the chance of survival of a patient across a given interval, given that they are alive at the start of that interval. Such an approach calls on a conditional probability structure. The conditional probabilities of survival across I_j, given alive at the start of I_j, are:

$$\begin{aligned} p_j &= P(\text{Surviving through } I_j | \text{ Alive at the start of } I_j) \\ &= P(Y > a_j | Y > a_{j-1}) \end{aligned}$$

$$= \frac{S(a_j)}{S(a_{j-1})}. \tag{5.1}$$

It follows that
$$p_1 = \frac{S(a_1)}{S(a_0)} = \frac{S(a_1)}{S(0)} = S(a_1)$$

since $S(0) = 1$. This means that the survival probabilities at the interval endpoints may be written as a telescoping product of terms:

$$S(a_j) = [S(a_1)][\frac{S(a_2)}{S(a_1)}][\frac{S(a_3)}{S(a_2)}]\ldots[\frac{S(a_{j-1})}{S(a_{j-2})}][\frac{S(a_j)}{S(a_{j-1})}] \tag{5.2}$$

and these telescoping terms on the right side of (5.2) are simply the conditional probabilities of survival determined by (5.1). This demonstrates that the probability of survival beyond an interval endpoint is a product of conditional probabilities as stated by Theorem 5.1.

Theorem 5.1 *In a lifetable built from intervals $I_j = [a_{j-1}, a_j)$, for $j = 1, 2, 3, \ldots, k+1$,*

$$S(a_j) = p_1 p_2 p_3 \ldots p_j,$$

where p_j is the conditional probability of survival across I_j given alive at the start of I_j. □

In lifetable estimation, we use Theorem 5.1 to estimate survival at the fixed interval endpoints.

For Step 2 in lifetable construction, clearly, the p_j need to be estimated so that S may be estimated at a_j. The most obvious estimator to use is the classical binomial estimator for a proportion:

$$\text{Estimate of } p_j = 1 - \frac{\text{number dying in } I_j}{\text{number with the potential to die in } I_j} \tag{5.3}$$

This estimate works because when estimating $1 - p_j$, we are estimating the proportion dying in the interval given that they were alive and under observation at the start of the interval. Therefore, as a simple binomial proportion, we estimate this as the number dying in the interval divided by the number at risk.

A problem arises with the estimation of the denominator of the fraction in (5.3) because potential deaths in the interval depend on the distribution of censored observations within the interval. Indeed, if all the censoring occurs immediately at the start of I_j, then the number at risk (with the potential of dying in the interval) is essentially $N_j - W_j$. On the other hand, if the censoring occurs just before the end of the interval, then the censored items were essentially at risk for the duration of the interval and the number at risk is N_j obtained by interpreting W_j as 0. For estimation purposes we average the two extremes in sample size,

$$\frac{(N_j - W_j) + (N_j)}{2} = N_j - \frac{1}{2}W_j,$$

and define the **effective number at risk**, N'_j, by

$$N'_j = N_j - \frac{1}{2}W_j.$$

This consideration of the effective number at risk in a given interval essentially assumes that the censoring occurs uniformly across the interval.

Following (5.3), and provided $N'_j > 0$, the **actuarial estimate** $\tilde{p}_j$ of p_j is taken as

$$\tilde{p}_j = 1 - \frac{D_j}{N'_j}, \quad j = 1, 2, 3, \ldots, k+1. \tag{5.4}$$

If, for a given interval, $N'_j = 0$, then the actuarial estimate of $\tilde{p}_j$ is taken as $\tilde{p}_j = 0$.

Finally for Step 3 in the construction of a lifetable,

$$\tilde{S}(a_j) = \tilde{p}_1 \tilde{p}_2 \tilde{p}_3 \ldots \tilde{p}_j, \quad j = 1, 2, 3, \ldots, k+1, \tag{5.5}$$

is the actuarial estimate of survival at $0 = a_0, a_1, a_2, \ldots, a_k, a_{k+1} = \infty$.

The product nature of the estimate in (5.5) arises from the product structure of $S(a_j)$ in Theorem 5.1. Clearly, if the jth interval lies beyond the support of observed lifetimes, there are no individuals at risk at the start of the interval, so $N_j = N'_j = 0$, $\tilde{p}_j = 0$ and $\tilde{S}(a_j) = 0$, clearly ensuring that the survival estimate reaches 0.

In deriving the estimate in (5.5), we have assumed that censored individuals have the same lifetime distributions as uncensored ones — p_j is the same for each individual — this may not be the same in a medical trial. Patients in a clinical trial may be lost to follow-up because of side effects experienced in their treatment group.

We can classify two basic assumptions for the construction of lifetables:

Assumption 1: The mechanism generating the censor times is independent of the observed lifetimes. This assumption is needed to ensure that p_j is the same for each individual.

Assumption 2: Both the death times (failure times) and the censor times in a given interval are assumed to be uniformly distributed across that interval. This assumption is necessary in the construction of the effective number at risk in a given interval.

The structure of a lifetable will vary across different types of software presentation. Essentially, we may regard a lifetable as being a generalisation of a frequency histogram to cater for right-censored data. The table will generally include listings of $I_j, n_j, d_j, w_j, \tilde{q}_j = 1 - \tilde{p}_j$, and $\tilde{S}(a_j)$. Some software provides more extensive listings (for example, the effective number at risk, or the hazard estimates in each interval). We demonstrate the structure, and how it may be determined by hand, in Example 5.6.

■ Example 5.6
Recursive calculations in a lifetable

The following data on 913 male and female patients with malignant melanoma, treated in the M. D. Anderson Tumor Clinic between 1944 and 1960 is from

Gross and Clark (1975). Here, survival time is defined as the time (years) from treatment to the death of the patient. For each j, let $\tilde{q}_j = 1 - \tilde{p}_j$.

I_j Interval (years)	d_j Deaths	w_j Losses	n_j Risk	n'_j Effective no. at risk	$\tilde{q}_j$	$\tilde{p}_j$	$\tilde{S}(a_j)$
$[0,1)$	312	96	913	865.0	0.361	0.639	0.639
$[1,2)$	96	74	505	468.0	0.205	0.795	0.508
$[2,3)$	45	62	335	304.0	0.148	0.852	0.433
$[3,4)$	29	30	228	213.0	0.136	0.864	0.374
$[4,5)$	7	40	169	149.0	0.047	0.953	0.356
$[5,6)$	9	37	122	103.5	0.087	0.913	0.325
$[6,7)$	3	17	76	67.5	0.044	0.956	0.311
$[7,8)$	1	12	56	50.0	0.020	0.980	0.305
$[8,9)$	3	8	43	39.0	0.077	0.923	0.281
$[9,\infty)$	32	-	32	32.0	1.000	0.000	0.000

Notice that if we examine the last two columns of the above table, and if we are careful with rounding errors, there is a demonstration of the multiplicative way that **lifetable estimates** may be determined recursively:

$$\tilde{S}(a_j) = \tilde{S}(a_{j-1})(1 - \frac{D_j}{N'_j}).$$

That is, from lines 1 and 2,

$$0.508 = \tilde{S}(a_2) = \tilde{p}_2 \tilde{S}(a_1) = (0.795)(0.639).$$

And more:

$$0.433 = \tilde{S}(a_3) = \tilde{p}_3 \tilde{S}(a_2) = (0.852)(0.508),$$

and so on.

The uniform choice of intervals does not seem particularly suitable since nearly half the patients died or are lost to follow-up in the first year. In cases such as this, medical checks may be taken annually following shorter check-up intervals post initial treatment. Assessments at three-month intervals in the first year would improve the accuracy of the table. To use the table, read the survival prospects for each year; for example, the lifetable estimate of the probability of a melanoma patient surviving more than 5 years after treatment is 0.356. □

5.5 Greenwood's Formula

It is unwise to use lifetable estimates $\tilde{S}(a_j)$ without a quotation of the standard error of the estimate. This is accomplished with the historical Greenwood's Formula, a famous result established a suprisingly long time ago (Greenwood,

1926). There have been many derivations of the standard error formula; we will concentrate on the derivation which approximates $\widetilde{S}(a_j)$ by a product of *independent* binomial proportions for the intervals of the lifetable prior to a_j. This will require positive sample sizes (effective numbers at risk) in each of the intervals concerned. If we condition on this, then the result is exact rather than an approximation; unconditionally, the standard error result is an approximation.

We begin by noting that if $N'_j > 0$, the effective number at risk, N'_j, for $j > r$ depends on past effective numbers at risk, N'_l with $l < r$ only through the value of N'_r. Of course, if $N'_j = 0$, then we cannot tell at a previous interval identified by $l < j$ whether $N'_l = 0$ or whether $N'_l > 0$. The discussion begins by showing that the lifetable estimate $\widetilde{S}(a_j)$ is 'approximately' unbiased for $S(a_j)$.

Theorem 5.2 $E[\widetilde{S}(a_j)] \approx p_1 p_2 p_3 \ldots p_j = S(a_j), \quad j = 1, 2, 3, \ldots, k+1.$

Proof: Let $i < j$. Then **conditional on** $N'_i > 0$, we have binomial counts for D_i:

$$D_i | N'_i \sim Bin(N'_i, 1 - p_i).$$

Using this result we find that

$$\begin{aligned}
E[1 - \widetilde{p}_i] &= E[\frac{D_i}{N'_i}] \\
&= E\{E[\frac{D_i}{N'_i} | N'_i]\} \\
&= E\{\frac{1}{N'_i} E[D_i | N'_i]\} \\
&= E\{\frac{1}{N'_i} [N'_i (1 - p_i)]\} \\
&= E\{(1 - p_i)\} \\
&= (1 - p_i)
\end{aligned}$$

Unconditionally, $E[1 - \widetilde{p}_i] \approx 1 - p_i$ and hence $E[\widetilde{p}_i] \approx p_i$. Similarly, by conditioning on $N'_i > 0$, it follows that for $l < i$, $E[\widetilde{p}_l \widetilde{p}_i] \approx p_l p_i$. In general, repeated conditional arguments show that

$$E[\widetilde{S}(a_j)] = E[\widetilde{p}_1 \widetilde{p}_2 \widetilde{p}_3 \ldots \widetilde{p}_j] \approx p_1 p_2 p_3 \ldots p_j = S(a_j).$$

□

The approximation is reasonably accurate provided $P(N'_j = 0)$ is small, as is generally the case.

The same method of taking repeated expected values is used to construct the proof of Greenwood's Formula for the standard error of the lifetable estimate.

Theorem 5.3 (Greenwood's Formula) *The standard error of the lifetable estimate is given by*

$$Var[\widetilde{S}(a_j)] \approx S(a_j)^2 \sum_{i=1}^{j} \frac{q_i}{p_i N_i'}, \quad j = 1, 2, 3, \ldots, k+1.$$

Proof: Let $i < j$. Then conditional on $N_i' > 0$, since the observed failures follow a binomial distribution,

$$D_i | N_i' \sim Bin(N_i', 1 - p_i),$$

it follows by the arguments of Theorem 5.2 that

$$Var[\widetilde{p}_i] = \frac{p_i(1 - p_i)}{N_i'},$$

so that, unconditionally, the second moment may be written as

$$E[\widetilde{p}_i^2] \approx \frac{p_i(1 - p_i)}{N_i'} + p_i^2 = p_i^2 \left(1 + \frac{q_i}{p_i N_i'}\right).$$

Combining these results gives

$$\begin{aligned} Var[\widetilde{S}(a_j)] &= E[\widetilde{S}(a_j)^2] - \{E[\widetilde{S}(a_j)]\}^2 \\ &\approx \prod_{i=1}^{j} p_i^2 \left(1 + \frac{q_i}{p_i N_i'}\right) - S(a_j)^2 \\ &= S(a_j)^2 \left[\prod_{i=1}^{j}\left(1 + \frac{q_i}{p_i N_i'}\right) - 1\right] \\ &\approx S(a_j)^2 \left[1 + \sum_{i=1}^{j} \frac{q_i}{p_i N_i'} - 1\right] \end{aligned}$$

where, in the last line, terms of order $\frac{1}{N_i'^2}$ are ignored when we expand

$$\prod_{i=1}^{j}(1 + a_i) \approx 1 + \sum_{i=1}^{j} a_i$$

for

$$a_i = \frac{q_i}{p_i N_i'}.$$

□

The estimation process of reporting $\widetilde{S}(a_j)$ to within two standard errors, where

$$S.E.[\widetilde{S}(a_j)] \approx \widetilde{S}(a_j) \sqrt{\sum_{i=1}^{j} \frac{\widetilde{q}_i}{\widetilde{p}_i N_i'}}, \quad j = 1, 2, 3, \ldots, k+1,$$

is reasonable provided the expected effective number at risk is not too small in any of the intervals.

Example 5.7
Standard error: 5-year survival prospects for melanoma patients
Following the lifetable in Example 5.6, we determine the appropriate standard error calculation for the confidence interval $\widetilde{S}(5) \pm 2S.E.[\widetilde{S}(5)]$ as

$$S.E.[\widetilde{S}(5)] \approx \widetilde{S}(5)\sqrt{\sum_{i=1}^{5} \frac{\widetilde{q}_i}{\widetilde{p}_i N'_i}}$$

$$= (0.356)\sqrt{\frac{0.047}{(0.953)(149)} + \frac{0.136}{(213)(0.864)} + \ldots + \frac{0.361}{(865)(0.639)}}$$

$$= 0.01899.$$

The realised confidence interval for $S(5)$ then reduces to 0.356 ± 0.038. □

Under heavier censoring patterns it is advisable to increase the number of intervals to smooth the effect that the increased censoring causes. Note also that there is no requirement to make all of the intervals the same length, although it may be convenient to do so.

Example 5.8
Time to breast cancer recurrence
The following is a MINITAB software analysis of a data set from Shouki and Pause (1999) for times (in months) until breast cancer recurrence in female patients. Much research has recently been directed at early detection through mass screening and an assessment of other variables that are predictive of remission times. In this study, information on other variables was also recorded. Only 17 observations were uncensored observed times until cancer recurrence and 56 observations were censored. C1= time to cancer recurrence and C2 holds the censor indicator for which $0 =$ no breast cancer recurrence and $1 =$ breast cancer recurrence. This is an example of a data set with a very high proportion of censoring, resulting in a high estimate of survival probability near the top of the distribution; note, for example, that $\widetilde{S}(120) = 0.6195$.

```
Distribution Analysis
  Variable: C1


Actuarial Table
```

Interval		Number	Number	Number	Conditional Probability	Standard
lower	upper	Entering	Failed	Censored	of Failure	Error
0.000000	50.0000	73	2	5	0.0284	0.0198
50.0000	70.0000	66	5	15	0.0855	0.0366
70.0000	80.0000	46	1	6	0.0233	0.0230
80.0000	90.0000	39	4	3	0.1067	0.0504
90.0000	100.0000	32	1	8	0.0357	0.0351
100.0000	120.0000	23	3	11	0.1714	0.0901
120.0000	140.0000	9	1	8	0.2000	0.1789

	Survival	Standard	95.0% Normal CI	
Time	Probability	Error	lower	upper
50.0000	0.9716	0.0198	0.9329	1.0000
70.0000	0.8886	0.0399	0.8105	0.9667
80.0000	0.8679	0.0440	0.7818	0.9541
90.0000	0.7753	0.0588	0.6601	0.8906
100.0000	0.7477	0.0629	0.6244	0.8709
120.0000	0.6195	0.0852	0.4526	0.7864
140.0000	0.4956	0.1301	0.2406	0.7505

MTB > print c1 c2

Data Display

Row	C1	C2	Row	C1	C2	Row	C1	c2
1	130	0	26	103	1	51	71	0
2	136	0	27	60	0	52	80	0
3	117	0	28	91	0	53	25	0
4	50	1	29	70	0	54	67	0
5	106	0	30	65	0	55	74	0
6	103	0	31	91	0	56	64	1
7	86	1	32	86	1	57	64	0
8	63	0	33	90	0	58	41	1
9	120	0	34	87	0	59	70	1
10	121	0	35	89	1	60	57	1
11	108	1	36	89	1	61	59	0
12	121	0	37	92	0	62	53	0
13	109	0	38	48	0	63	69	0
14	111	0	39	89	0	64	55	0
15	60	1	40	95	0	65	58	1
16	106	0	41	91	0	66	68	0
17	108	0	42	47	1	67	60	0
18	105	0	43	75	0	68	126	1
19	98	0	44	49	0	69	127	0
20	108	1	45	66	0	70	126	0
21	62	0	46	65	0	71	102	0
22	106	0	47	22	0	72	122	0
23	95	1	48	73	0	73	100	0
24	94	0	49	67	0			
25	19	0	50	75	0			

5.6 Exercises

5.1. For the Western hostages data of Example 5.1, what is the response variable of interest? Discuss the presence of left- or right-censoring in the data.

Construct the data set of 'observed months of captivity for Western hostages' by interpreting the diagrams in the newspaper report. Indicate using a '+' which data points are right-censored. (For example, 20+ is a right-censored observation of 20 months.)

5.2. De Stavola and Christensen (1996) present a lifetable (times measured in years) summarising a clinical trial, called the PCB1 trial, for the treatment of primary biliary cirrhosis, a potentially fatal liver condition.

Interval (years)	Total	Deaths	Losses
[0, 1)	191	18	14
[1, 2)	159	19	9
[2, 3)	131	13	9
[3, 4)	109	9	12
[4, 5)	88	14	7
[5, 6)	67	7	14
[6, 7)	46	10	12
[7, 8)	24	3	4
[8, 9)	17	2	3
[9, 10)	12	1	6
[10, 11)	5	0	3
[11, 12)	2	0	2

(*Source*: De Stavola, B. L. and Christensen, E. (1996). Multi-level models for longitudinal variables prognostic for survival. *Lifetime Data Analysis*, 2, 329–47.)

(a) Construct further columns of this lifetable giving the standard quantities n'_j, $\tilde{q}_j$, $\tilde{p}_j$ and $\tilde{S}(a_j)$.
(b) Use the lifetable to estimate the probability of survival of PCB patients beyond 5 years.
(c) For patients who have survived 5 years, estimate the probability of surviving a further year.

5.3. (Continuation). Recall from equation (1.11) the formula derived for the conditional mean survival, for example, for patients who have lived at least 2 years, as
$$E(Y|Y > 2) = 2 + \frac{\int_2^\infty S(y)dy}{S(2)}.$$
Discuss how this quantity might be estimated using information from the lifetable for the PCB1 trial.

5.4. Collett (1994) reports data for the time to recurrence of an ulcer after an initial ulcer has been diagnosed, treated and healed. The data arose from six monthly clinic visits where endoscopies were performed: a positive test result (result = 2) indicates that the remission ceased in the time interval since the last scheduled visit. A negative test result (result =1) indicates that the patient is still in remission. Further, endoscopies were performed at in-between times for concerned patients presenting with symptoms. Such patients with positive test results provide 'exact' observed remission times. Also recorded was the patient's age (years) at entry into the program and treatment type **Tt** denoted by A or B.

Construct the censoring intervals, I_i for $i = 1, 2, 3, \ldots, 42$, for the 42 patients listed. Clearly indicate which patients are right-censored, which are left-censored and which are interval-censored.

Age	Tt	Prior visit	Result	Age	Tt	Prior visit	Result
48	B	7	2	63	B	12	1
73	B	12	1	41	A	12	1
54	B	12	1	47	B	12	1
58	B	12	1	58	A	3	2
56	A	12	1	74	A	2	2
49	A	12	1	75	A	6	1
71	B	12	1	72	A	12	1
41	A	12	1	59	B	12	2
23	B	12	1	52	B	12	1
37	B	5	2	75	B	12	2
38	B	12	1	76	A	12	1
76	B	12	1	34	A	6	1
38	A	12	1	36	B	12	1
27	A	6	2	59	B	12	1
47	B	6	2	44	A	12	2
54	A	6	1	28	B	12	1
38	B	10	2	62	B	12	1
27	B	7	2	23	A	12	1
58	A	12	1	49	B	12	1
75	B	12	1	61	A	12	1
25	A	12	1	33	B	12	1

(*Source*: Collett, D. (1994). *Modelling Survival Data in Medical Research.* Chapman & Hall: London.)

5.5. (**Time to breast cancer recurrence.**) Example 5.8 showed a MINITAB software lifetable analysis of a data set from Shouki and Pause (1999) for times Y (in months) until breast cancer recurrence in female patients. Use this output to answer the following:

EXERCISES

(a) Explain which of the following terms apply to these data: Type I censoring; Type II censoring; right-censored; interval-censored.
(b) Determine the effective number at risk for the time interval $[80, 90]$ months.
(c) Use the lifetable to:
   (i.) Estimate the probability of patient survival beyond 100 months.
   (ii.) Estimate the probability of patient survival beyond 100 months given patient alive at 90 months.
   (iii.) Estimate $E(Y|Y > 80)$.

5.6. Discuss the type of censoring evident in the Stanford Heart Transplant data. Explain how this data differs from truncated data.

5.7. Construct a lifetable for the Stanford Heart Transplant data. Use the lifetable to estimate the probability of surviving a heart transplant beyond 1000 days. Give the standard error of your estimate.

5.8. Show that left-censoring is a special case of right-censoring with a reversed timescale.

5.9. Construct a formula for $E(Y|Y > 2)$ when a survival random variable Y with hazard function $h(u) = 2$ is right-censored at $u = 2$. [*Hint*: Use the results of equation (1.11).]

Note that $E(Y|Y > t)$ gives the mean lifetime, given that right-censoring has occurred at t.

5.10. Construct a formula for $E(Y|Y < l)$ when a survival random variable Y with survival function S is left-censored by a fixed censor time l. [*Hint*: Use the results of Chapter 1.]

Note that $E(Y|Y < l)$ gives the mean lifetime, given that left-censoring has occurred at l.

5.11. Construct a formula for $E(Y|l < Y < u)$ when a survival random variable Y with survival function S is interval-censored by a fixed censoring interval $[l, u]$.

(This situation may occur for grouped data where patients are checked at regular time intervals for the presence of, for example, breast cancer. This would give the mean time to onset of the disease if the disease was absent at u, but present by l. See, for an analysis of doubly censored grouped data, Turnbull, 1974.)

5.12. According to Definition 2.9, the Pareto distribution with $\alpha > 0$ and $\gamma > 0$ has hazard function
$$h(y) = \frac{\alpha}{y} \text{ for } y \geq \gamma.$$

(a) Determine the Pareto survival function and density function.
(b) Show that if $Y \sim Pareto(\alpha, \gamma)$, then the conditional distribution of Y given $Y > t$ is also a Pareto distribution, and give the parameters of that distribution.

(c) What are the implications for left-truncated Pareto data?

5.13. (**Probability plots for Type II censored data.**) Lawless (1982) considers probability plots for the Mann and Fertig (1973) data, discussed in this chapter, on the failure times of aircraft components. The data were obtained from $n = 13$ randomly selected test items and the lifetest terminated at the observed failure of the 10th item. The $r = 10$ observed lifetimes were (in hours)

$$0.22, \ 0.50, \ 0.88, \ 1.00, \ 1.32, \ 1.33, \ 1.54, \ 1.76, \ 2.50, \ 3.00.$$

Construct a Weibull probability plot for these data using $n = 13$ in the determination of the *plotting positions*, but plot only the ten observed lifetimes. What conclusions may be drawn from this probability plot?

5.14. Use the definition of $\widetilde{S}(a_j)$ to show that the estimates in a lifetable may be determined recursively by the formula

$$\widetilde{S}(a_j) = \widetilde{S}(a_{j-1})(1 - \frac{D_j}{N'_j}), \quad j = 1, 2, 3, \ldots, k+1.$$

5.15. Show that if there is no censoring in the lifetable, then

$$\widetilde{p}_i = \frac{N_{i+1}}{N_i}, \quad i = 1, 2, 3, \ldots, k+1.$$

Hence show that $\widetilde{S}(a_j) = \frac{N_{j+1}}{n}$.

5.16. Show that if there is no censoring in the lifetable, $N'_i \widetilde{p}_i = n\widetilde{S}(a_i)$. Substitute this result into Greenwood's Formula to obtain

$$\{S.E.[\widetilde{S}(a_j)]\}^2 \approx \frac{\widetilde{S}(a_j)^2}{n} \sum_{i=1}^{j} (\frac{1}{\widetilde{S}(a_i)} - \frac{1}{\widetilde{S}(a_{i-1})}), \quad j = 1, 2, 3, \ldots, k+1.$$

Simplify the right-side of this expression to show that Greenwood's Formula reduces to the usual estimate for a binomial proportion:

$$\{S.E.[\widetilde{S}(a_j)]\}^2 \approx \frac{\widetilde{S}(a_j)(1 - \widetilde{S}(a_j))}{n}, \quad j = 1, 2, 3, \ldots, k+1.$$

5.17. (**Alternative derivation of Greenwood's Formula.**) Use the Delta Method of Theorem 4.1 to determine

$$Var[\log_e \widetilde{p}_j], \quad j = 1, 2, 3, \ldots, k+1.$$

Hence, assuming $\log_e \widetilde{p}_1, \log_e \widetilde{p}_2, \ldots, \log_e \widetilde{p}_j$ to be independent, show that

$$S.E.[\log_e \widetilde{S}(a_j)] \approx \sqrt{\sum_{i=1}^{j} \frac{\widetilde{q}_i}{\widetilde{p}_i N'_i}}, \quad j = 1, 2, 3, \ldots, k+1.$$

EXERCISES

Hence use the Delta Method again to establish Greenwood's Formula for
$$S.E.[\tilde{S}(a_j)].$$

5.18. The data in Example 5.3 from the Anti-Cancer Council of Victoria show survival time in months from date of diagnosis of Astrocytomas (brain tumors) until death resulting from the effects of tumors.

   (a) Rank these data from smallest to largest. Does the censoring appear uniform across the support of the lifetime distribution? Explain briefly.
   (b) Analyse these data in a lifetable.
   (c) Estimate the probability of survival for more than a year after Astrocytomas diagnosis. Give the standard error of estimate.

5.19. (**Hazard estimation in the lifetable.**) Recall from Example 2.1 the simple method for hazard estimation
$$\frac{P(y < Y < y + \Delta y)}{(\Delta y) P(Y > y)}$$
for estimating the average failure rate across the interval $[y, y + \Delta y)$. If we apply this to the interval $I_j = [a_{j-1}, a_j)$, the hazard estimate may be estimated at the midpoint m_j of I_j provided I_j is of finite width (i.e., $j \leq k$.)

   (a) Show, using linear interpolation, that $\tilde{S}(m_j)$ may be found by solving the equation
$$\frac{\tilde{S}(m_j) - \tilde{S}(a_j)}{\tilde{S}(a_{j-1}) - \tilde{S}(a_j)} = \frac{1}{2}.$$
   (b) Hence show that the hazard estimate at m_j is given by
$$\tilde{h}(m_j) = \frac{2[\tilde{S}(a_{j-1}) - \tilde{S}(a_j)]}{[a_j - a_{j-1}][\tilde{S}(a_{j-1}) + \tilde{S}(a_j)]}.$$
   (c) Use (b) to estimate the hazard rate at the midpoints of the intervals of the lifetable for patients with malignant melanomas in Example 5.6.

5.20. Example 5.4 presents a left-censored data set of times to event 'half the baboon troop descends from the trees' as records on a 24-hour clock. By subtracting the observed times from 2400 hours (midnight), so that the transformed times are represented as minutes until midnight that the descent occurs, show that the data set becomes right-censored.

Discuss how a lifetable may be used to estimate the average time at which half the troop has descended.

CHAPTER 6

The Product-Limit Estimator

The **product-limit estimator**, developed by Kaplan and Meier (1958), estimates the survival function, S, using right-censored data. It has a similar definition to the actuarial estimator, except that the interval endpoints are no longer fixed; the intervals comprise the spaces between the data as they are observed. Some of the data may be censored and some of the data may be uncensored.

6.1 Fitting the estimator

Suppose that $Y_1, Y_2, \ldots, Y_n$ are **right-censored** by constants $t_1, t_2, \ldots, t_n$. We observe (Z_i, δ_i), $i = 1, 2, \ldots, n$, where $Z_i = \min\{Y_i, t_i\}$ and

$$\delta_i = \begin{cases} 1 & \text{if } Y_i \leq t_i \quad \text{(uncensored)} \\ 0 & \text{if } Y_i > t_i \quad \text{(censored)}. \end{cases}$$

Because the data occur naturally as ordered pairs, we know exactly which points are censored and which are not. Assume initially that there are no tied observations, so that $Z_{(1)} < Z_{(2)} < Z_{(3)} < \ldots < Z_{(n)}$ and the observed data are all different. We always have to ensure that when the data are ranked from smallest to largest, that the indicators are rearranged to follow the new ranking; that is, we let $\delta_{(i)}$ denote the value of the indicator associated with $Z_{(i)}$. In fact, $\delta_{(i)} = \delta_j$ when $Z_{(i)} = Z_j$, so that $\delta_{(1)}, \delta_{(2)}, \delta_{(3)}, \ldots, \delta_{(n)}$ themselves are not ordered. As was the case with lifetable estimation, we assume that the survival variable being measured is independent of the mechanism generating the censor times.

To construct the survival estimator, divide $(0, Z_{(n)})$, the support of the observed lifetime distribution, into disjoint intervals

$$I_j = (Z_{(j-1)}, Z_{(j)}], \quad j = 1, 2, 3, \ldots, n, \quad \text{such that} \quad Z_0 = 0.$$

Definition 6.1 *The **risk set at time** u, denoted by $R(u)$, is the set of indices of subjects which are still alive and under observation at time u^-, that is, at time just prior to u.* □

Then:
N_j, the **number at risk**, is the number of elements in the set $R(Z_{(j)})$;
D_j, the **number of deaths**, is the observed failures at $Z_{(j)}$ (and is 0 or 1);
p_j is the conditional probability $P(\text{Surviving through } I_j | \text{ Alive at the start of } I_j)$.

N_j and the risk set are used to estimate p_j. Clearly, for estimation purposes,

$$\text{Estimate of } p_j = 1 - \frac{\text{number dying in } I_j}{\text{number with the potential to die in } I_j}$$

$$= \begin{cases} 1 - \frac{1}{N_j} & \text{if } \delta_{(j)} = 1; \\ 0 & \text{if } \delta_{(j)} = 0. \end{cases}$$

The product of all such estimates, $\widehat{p}_j$, gives an estimate of the survival function S. Note that as j increases, the risk sets diminish one at a time: $N_j = n - (j-1) = n - j + 1$. Putting these ideas together, for a fixed value of u,

$$\widehat{S}(u) = \prod_{j:Z_{(j)} \leq u} \widehat{p}_j$$

$$= \prod_{j:Z_{(j)} \leq u, \delta_{(j)} = 1} (1 - \frac{1}{N_j})$$

$$= \prod_{j:Z_{(j)} \leq u} (1 - \frac{1}{N_j})^{\delta_{(j)}}$$

$$= \prod_{j:Z_{(j)} \leq u} (1 - \frac{1}{n-j+1})^{\delta_{(j)}}$$

$$= \prod_{j:Z_{(j)} \leq u} (\frac{n-j}{n-j+1})^{\delta_{(j)}}.$$

Definition 6.2 *If u is fixed, and there are no ties in the observed possibly right-censored data*

$$(Z_{(1)}, \delta_{(1)}), (Z_{(2)}, \delta_{(2)}), \ldots, (Z_{(n)}, \delta_{(n)}),$$

then the **Kaplan-Meier product-limit estimator** $\widehat{S}$ *of S (often abbreviated to* **PL-estimator***) is defined by*

$$\widehat{S}(u) = \prod_{j:Z_{(j)} \leq u} (\frac{n-j}{n-j+1})^{\delta_{(j)}}.$$

□

We shall see in Theorem 6.1 some basic characteristics of the behaviour of $\widehat{S}(u)$ for $-\infty < u < \infty$. To assist us in understanding the shape of $\widehat{S}(u)$ at this point in the discussion, we can see immediately see by examining the formula for $\widehat{S}(u)$ in Definition 6.2 that:

1. $\widehat{S}(0) = 1$;
2. $\widehat{S}(u)$ is monotone decreasing;
3. $\widehat{S}(u)$ is piecewise constant;

FITTING THE ESTIMATOR

4. $\widehat{S}(u)$ has jump discontinuities at the uncensored observations;

5. $\widehat{S}(u)$ has no jump discontinuities at the censored observations.

We leave the size of the jump discontinuities to the proof of Theorem 6.1.

The process of fitting the product-limit estimator to a given set of survival data will involve certain conventions, modifications and implementations which we label as **Implementations 1, 2, and 3**. These conventions allow us to handle tied data, where the ties may be between uncensored observations, and ties between censored and uncensored observations. The conventions also apply to situations where the largest observation is censored; because $\widehat{S}$ jumps only at the uncensored observations, this would leave $\widehat{S}(u)$ hanging above 0 for large values of u and unable to eventually reach the estimation target 0.

> **Implementation 1: Uncensored ties.** Suppose that just before time u, there are m individuals alive and at u there are d deaths (uncensored ties); this corresponds to a factor $(1 - \frac{d}{m})$ in the PL-estimator.

Implementation 1 is to resolve ties between uncensored values. To do this we imagine the d deaths split across an infinitesimally small time frame. This simple resolution then creates the following 'factor' Δ in the PL-estimator, which we can see is composed of d telescoping factors, one for each death:

$$\begin{aligned}
\Delta &= (1 - \frac{1}{m})(1 - \frac{1}{m-1})(1 - \frac{1}{m-2})\ldots(1 - \frac{1}{m-(d-1)}) \\
&= (\frac{m-1}{m})(\frac{m-2}{m-1})(\frac{m-3}{m-2})\ldots(\frac{m-d}{m-d+1}) \\
&= (\frac{m-d}{m}) \\
&= (1 - \frac{d}{m}).
\end{aligned}$$

> **Implementation 2: Censored/uncensored ties.** Uncensored observations precede censored observations with which they are tied.

It is a natural justice that guides Implementation 2. Censored observations which are tied with uncensored observations are 'untied' by remembering that right-censored observations would be larger than observed had there been no censoring. In this case it is sensible that uncensored observations precede censored observations with which they are tied.

> **Implementation 3:** Always set $\delta_{(n)} = 1$. That is, the largest observation always has its status (re-)assigned as uncensored.

This last simple convention was pioneered by Efron (1967) and Miller (1976,

1981). This has the effect of ensuring that the PL-estimator finally reaches 0. That is, $\widehat{S}(u) = 0$ for $u > Z_{(n)}$, so that $\widehat{S}$ eventually reaches 0 just as S does.

These implementations allow Definition 6.2 of the PL-estimator to be generalised to account for tied observations.

Definition 6.3 *If u is fixed, and the distinct right-censored data are denoted by*
$$(Z'_{(1)}, \delta'_{(1)}), (Z'_{(2)}, \delta'_{(2)}), \ldots, (Z'_{(k)}, \delta'_{(k)}),$$
*and D_j deaths occur at $Z'_{(j)}$, then the **Kaplan-Meier product-limit estimator** $\widehat{S}$ of S is defined by*
$$\widehat{S}(u) = \prod_{j: Z'_{(j)} \le u} (1 - \frac{D_j}{N_j})^{\delta'_{(j)}}.$$

□

We are now in a position to examine the characteristics of the graph of $\widehat{S}(u)$ against u.

Theorem 6.1 $\widehat{S}$ *is a piecewise constant function of u which is right-continuous with discontinuities (or jump points) occurring at the uncensored observations. The size of the discontinuity at $u = z_{(j)}$ when $\delta_{(j)} = 1$ is*
$$\frac{\widehat{S}(z_{(j-1)})}{n - j + 1}$$
when there are no ties at $z_{(j)}$.

Proof: Suppose that $\delta_{(j)} = 1$. The definition
$$\widehat{S}(u) = \prod_{j: z_{(j)} \le u} (\frac{n - j}{n - j + 1})^{\delta_{(j)}}$$
implies that
$$\widehat{S}(z_{(j)}^+) = \lim_{\epsilon \downarrow 0} \widehat{S}(z_{(j)} + \epsilon) = \lim_{\epsilon \downarrow 0} \prod_{i: z_{(i)} \le z_{(j)} + \epsilon} (\frac{n - i}{n - i + 1})^{\delta_{(i)}} = \widehat{S}(z_{(j)}).$$
Similarly,
$$\widehat{S}(z_{(j)}^-) = \lim_{\epsilon \downarrow 0} \widehat{S}(z_{(j)} - \epsilon) = \lim_{\epsilon \downarrow 0} \prod_{i: z_{(i)} \le z_{(j)} - \epsilon} (\frac{n - i}{n - i + 1})^{\delta_{(i)}}$$
which, by the definition of $\widehat{S}$, may be simplified to
$$\lim_{\epsilon \downarrow 0} \prod_{i: z_{(i)} \le z_{(j-1)}} (\frac{n - i}{n - i + 1})^{\delta_{(i)}} = \prod_{i: z_{(i)} \le z_{(j-1)}} (\frac{n - i}{n - i + 1})^{\delta_{(i)}} = \widehat{S}(z_{(j-1)}).$$
This shows that $\widehat{S}$ is continuous from the right. The size of the discontinuity

FITTING THE ESTIMATOR

at $u = z_{(j)}$ when $\delta_{(j)} = 1$ is (when there are no ties at $z_{(j)}$)

$$\begin{aligned}\widehat{S}(z_{(j)}^-) - \widehat{S}(z_{(j)}^+) &= \widehat{S}(z_{(j-1)}) - \widehat{S}(z_{(j)}) \\ &= \widehat{S}(z_{(j-1)}) - \left(\frac{n-j}{n-j+1}\right)\widehat{S}(z_{(j-1)}) \\ &= \widehat{S}(z_{(j-1)})\left[1 - \left(\frac{n-j}{n-j+1}\right)\right] \\ &= \frac{\widehat{S}(z_{(j-1)})}{n-j+1}.\end{aligned}$$

□

The obvious generalisation to Theorem 6.1 occurs when there are d_j uncensored ties at $z'_{(j)}$. They collectively each add equally to the size of the jump. The size of the discontinuity at $u = z'_{(j)}$ when $\delta'_{(j)} = 1$ is

$$\widehat{S}(z'_{(j-1)})\frac{d_j}{n_j}$$

when there are d_j uncensored ties at $z_{(j)}$.

Having determined the jump size occurring at the uncensored points, it follows by the above arguments that the PL-estimator may be determined recursively at adjacent uncensored points, say z'_{j-1} and z'_j: once $\widehat{S}(z'_{(j-1)})$ is known, then $\widehat{S}(z'_{(j)})$ follows from the recursive relationship

$$\widehat{S}(z'_{(j)}) = \widehat{S}(z'_{(j-1)})\left(1 - \frac{d_j}{n_j}\right). \tag{6.1}$$

This will be demonstrated in Example 6.1.

Before performing detailed calculations of the PL-estimator, consider what occurs to $\widehat{S}$ when there is no censoring present: when there are no ties, the jump sizes all revert to being the same size $\frac{1}{n}$. We leave the proof of this conclusion to the exercises.

Theorem 6.2 *When there is no censoring present in the observed data, the PL-estimator $\widehat{S}(u)$ reverts to the empirical survivor function*

$$S_n(u) = \frac{\text{number of observations} > u}{n}$$

for estimating $S(u)$.

□

■ Example 6.1
Patient monitoring after Astrocytomas diagnosis: constructing the Product-Limit estimator under heavy right-censoring

The data of Example 5.3 from the Anti Cancer Council of Victoria show survival time in months from date of diagnosis of Astrocytomas (brain tumors) until death resulting from the effects of tumors. We now examine the hand

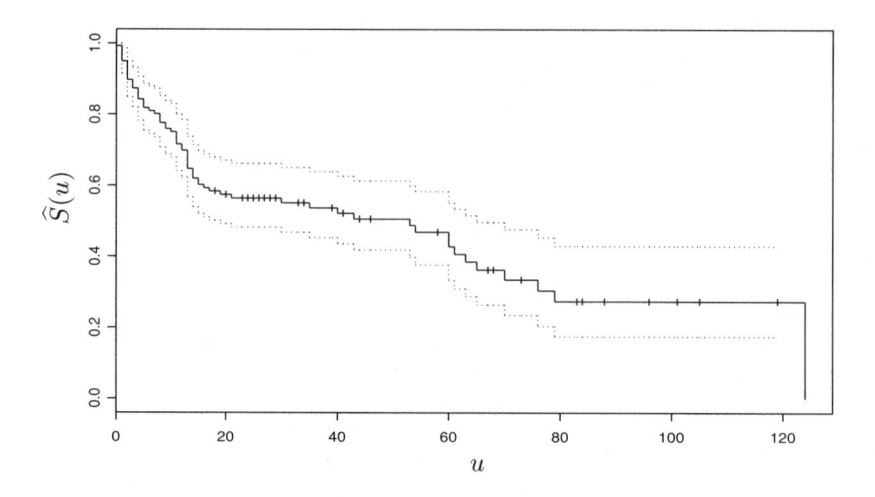

Figure 6.1 *The PL-estimator for the survival time (months) after Astrocytomas diagnosis.*

calculation, software calculation and graphical calculation of the PL-estimator of the survival function. The heavy censoring pattern (approximately 50%) especially at the top of the distribution, makes long-term survival prospects difficult to ascertain without the assistance of the PL-estimator.

The standard + notation represents right-censoring caused by withdrawal from the monitoring process, or a recording of the current status as 'alive', or even death from other causes. Almost half the subjects were in this last category as patients were admitted into the study at different stages depending on the date of original diagnosis. These data were part of a larger ongoing study of brain tumor patients in Victoria and New South Wales.

Using S-PLUS software, the graph of the PL-estimator is given by the dark lines in Figure 6.1, with the lighter lines delimiting a confidence band constructed using Greenwood's Formula as described in Section 6.3. It is noticeable that the confidence band grows wider toward the top of the survival distribution as it is based on ever-decreasing numbers of censored and uncensored observations. Tick marks optionally included on the dark lines denote the location of censored points. The plotting procedure evident in S-PLUS is to draw a continuous dark monotone decreasing line rather than the right-continuous step function dictated by the theory of Theorem 6.1. With a lot of observations plotted, it is the shape of the resulting 'curve' which becomes more important. In this case, the longevity of many patients is shown by the flattening of the curve after the initial steep decline. The large jump at the largest uncensored observation is quite noticeable. The graph of the PL-

FITTING THE ESTIMATOR

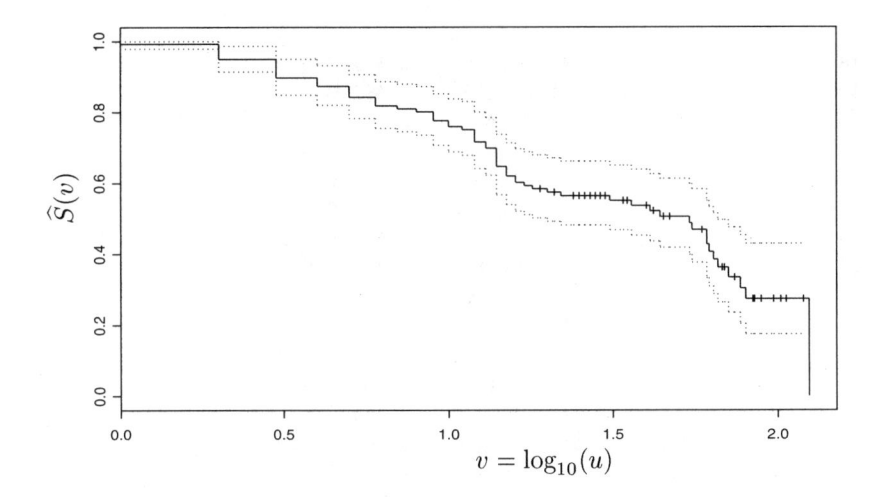

Figure 6.2 *The PL-estimator for the survival time $v = \log_{10}(u)$ (in log base 10 months) after Astrocytomas diagnosis.*

estimator is also presented on the log base 10 scale in Figure 6.2, a scale often appropriate for the presentation of lifetime data.

Edited software output from fitting the `survfit` function in S-PLUS is as follows:

```
*** Nonparametric Survival ***

  n   events  mean  se(mean)  median  0.95LCL  0.95UCL
 140    68     55     5.25      53      19       65

time  n.risk  n.event  survival  std.err      95% CI
  0    140       1      0.993   0.00712    0.979  1.000
  1    139       6      0.950   0.01842    0.915  0.987
  2    127       7      0.898   0.02594    0.848  0.950
  3    114       3      0.874   0.02862    0.820  0.932
  4    110       4      0.842   0.03169    0.782  0.907
  5    104       3      0.818   0.03373    0.754  0.887
  6     99       1      0.810   0.03439    0.745  0.880
  7     96       1      0.801   0.03505    0.735  0.873
  8     95       3      0.776   0.03686    0.707  0.852
  9     92       2      0.759   0.03794    0.688  0.837
 10     89       1      0.751   0.03846    0.679  0.830
 11     87       4      0.716   0.04038    0.641  0.800
 12     83       2      0.699   0.04121    0.623  0.784
 13     81       6      0.647   0.04324    0.568  0.738
 14     73       3      0.620   0.04410    0.540  0.713
```

15	69	2	0.602	0.04462	0.521	0.697
16	66	1	0.593	0.04487	0.512	0.688
17	65	1	0.584	0.04509	0.502	0.680
18	64	0	0.584	0.04509	0.502	0.680
19	62	1	0.575	0.04534	0.492	0.671
20	59	0	0.575	0.04534	0.492	0.671
21	57	1	0.565	0.04565	0.482	0.662
23	56	0	0.565	0.04565	0.482	0.662
24	55	0	0.565	0.04565	0.482	0.662
25	52	0	0.565	0.04565	0.482	0.662
26	51	0	0.565	0.04565	0.482	0.662
27	47	0	0.565	0.04565	0.482	0.662
28	45	0	0.565	0.04565	0.482	0.662
29	44	0	0.565	0.04565	0.482	0.662
30	43	1	0.552	0.04644	0.468	0.651
33	41	0	0.552	0.04644	0.468	0.651
34	40	0	0.552	0.04644	0.468	0.651
35	39	1	0.537	0.04735	0.452	0.639
39	37	0	0.537	0.04735	0.452	0.639
40	36	1	0.522	0.04834	0.436	0.626
41	33	0	0.522	0.04834	0.436	0.626
43	32	1	0.506	0.04951	0.418	0.613
44	31	0	0.506	0.04951	0.418	0.613
46	30	0	0.506	0.04951	0.418	0.613
53	28	1	0.488	0.05093	0.398	0.599
54	25	1	0.469	0.05250	0.376	0.584
58	24	0	0.469	0.05250	0.376	0.584
60	23	2	0.428	0.05528	0.332	0.551
61	20	1	0.406	0.05650	0.309	0.534
63	19	1	0.385	0.05744	0.287	0.516
65	17	1	0.362	0.05835	0.264	0.497
67	15	0	0.362	0.05835	0.264	0.497
68	14	0	0.362	0.05835	0.264	0.497
70	13	1	0.335	0.06015	0.235	0.476
73	12	0	0.335	0.06015	0.235	0.476
76	11	1	0.304	0.06190	0.204	0.453
79	10	1	0.274	0.06273	0.175	0.429
83	9	0	0.274	0.06273	0.175	0.429
84	7	0	0.274	0.06273	0.175	0.429
88	6	0	0.274	0.06273	0.175	0.429
96	5	0	0.274	0.06273	0.175	0.429
101	4	0	0.274	0.06273	0.175	0.429
105	3	0	0.274	0.06273	0.175	0.429
119	2	0	0.274	0.06273	0.175	0.429
124	1	1	0.000	NA	NA	NA

Hand-calculations which yield the figures listed in the edited software output are given for the lowest censored and uncensored data points at the left-hand end of the distribution (when most patients are at risk). This assists

SELF-CONSISTENCY

in understanding the software output both in result and structure. The calculations below demonstrate how the PL-estimator may be determined at consecutive uncensored points by repeated use of a multiplying factor as is detailed in equation (6.1). Because the one observed death at time 0 is likely the valid result of rounding down from a fraction of a month, say 0^+, $\widehat{S}(0^+)$ is set at $(1 - \frac{1}{140}) = 0.993$ rather than $\widehat{S}(0) = 1$. From this starting point, using (6.1), we obtain

$$\widehat{S}(0^+) = 0.993$$
$$\widehat{S}(1) = \widehat{S}(0^+)(1 - \frac{6}{139}) = 0.950$$
$$\widehat{S}(2) = \widehat{S}(1)(1 - \frac{7}{127}) = 0.898$$
$$\widehat{S}(3) = \widehat{S}(2)(1 - \frac{3}{114}) = 0.874$$
$$\widehat{S}(4) = \widehat{S}(3)(1 - \frac{4}{110}) = 0.842$$
$$\widehat{S}(5) = \widehat{S}(4)(1 - \frac{3}{104}) = 0.818$$
$$\widehat{S}(6) = \widehat{S}(5)(1 - \frac{1}{99}) = 0.810$$
$$\widehat{S}(7) = \widehat{S}(6)(1 - \frac{1}{96}) = 0.801$$
$$\widehat{S}(8) = \widehat{S}(7)(1 - \frac{3}{95}) = 0.776$$
$$\vdots = \vdots$$

These calculations continue in the manner listed in the extensive software output. □

6.2 Self-consistency

Intuitively, an estimator S^* of S is called **self-consistent** if at any point u on the time axis, $S^*(u)$ incorporates in a consistent way the information supplied by S^* at *censored* points earlier on the time axis.

Because of the right-censoring, at any censored point $Z_{(i)} < u$, we can only estimate whether or not the corresponding true lifetime $Y_{(i)}$ exceeds u. For example, in a clinical trial, if a patient moves away to another city and is lost to follow-up, their survival time can only be estimated based on the fact that they were fine at the last time of contact. If we regard the point u on the time axis as 'now', then the chance of a past patient who is lost to follow-up at time $z_{(i)}$ still being alive has probability

$$P(Y \geq u | Y \geq z_{(i)}) = \frac{S(u)}{S(z_{(i)})}$$

of occurring, assuming that the mechanism generating the censor times and the survival variables are independent. This probability is naturally estimated by
$$\frac{S^*(u)}{S^*(z_{(i)})}.$$
To be self-consistent, S^* counts the number of points which exceed u and also includes in the count estimates of those points which *would* exceed u had the deaths taken a natural course and patients not been lost to follow-up.

Paralleling the definition
$$S_n(u) = \frac{\text{number of observations} > u}{n} = \frac{1}{n} \sum_{i=1}^{n} I_{(u,\infty)}(Y_i)$$
of the empirical survivor function from Definition 1.3, we have:

Definition 6.4 *An estimator S^* of S is **self-consistent** if it satisfies the equation*
$$S^*(u) = \frac{1}{n} \sum_{i=1}^{n} \left[I_{(u,\infty)}(Z_{(i)}) + \frac{S^*(u)}{S^*(z_{(i)})} I_{(-\infty,u)}(Z_{(i)})(1 - \delta_{(i)}) \right] \quad (6.2)$$
for all values of u. □

Efron (1967) showed that there is a unique solution to (6.2) and it is the Kaplan-Meier product-limit estimator.

Theorem 6.3 *For all values of $u > 0$, $S^*(u) = \widehat{S}(u)$.* □

This result is also established in Miller (1981) by showing that the points of discontinuity for each of the piecewise constant estimators S^* and $\widehat{S}$ occur in the same places and the jumps at the points of discontinuity are of the same magnitude. Further, (6.2), when used iteratively, is able to construct the PL-estimator from the biased estimator
$$S^{(0)}(u) = \frac{1}{n} \sum_{i=1}^{n} I_{(u,\infty)}(Z_{(i)}).$$
This is done by constructing a convergent sequence of estimators
$$S^{(0)}, S^{(1)}, S^{(3)}, S^{(4)}, \ldots$$
through repeated substitution into
$$S^{(j+1)}(u) = \frac{1}{n} \sum_{i=1}^{n} \left[I_{(u,\infty)}(Z_{(i)}) + \frac{S^{(j)}(u)}{S^{(j)}(z_{(i)})} I_{(-\infty,u)}(Z_{(i)})(1 - \delta_{(i)}) \right]$$
as suggested by (6.2). Again the sequence of estimators converges to the PL-estimator. This is the approach taken by Turnbull (1974, 1976) to construct self-consistent estimators under more complex double-censoring patterns.

Further to the historical underpinning of the structure of the PL-estimator, we note finally that $\widehat{S}$ is the nonparametric maximum likelihood estimator of S. This is eloquently derived, albeit under a different system of notation in Kalbfleisch and Prentice (1980), page 10. We defer full discussion of maximum likelihood methods when fitting (parametric) models to censored data until the next chapter.

6.3 Standard errors

Meier (1975) considers the evolution of $\widehat{S}(u)$ to be a stochastic process in u. By also considering the stochastic nature of $N(u)$, the number of individuals in the risk set at time u, as time passes, it may be established that $\widehat{S}(u)$ is asymptotically normally distributed. The importance of this is that it allows $S(u)$ to be estimated at time u by the point estimator $\widehat{S}(u)$ and by the approximate confidence interval

$$\widehat{S}(u) \pm 2 S.E.[\widehat{S}(u)]$$

based on two standard errors.

For finite sample sizes, Meier (1975) shows that $\widehat{S}(u)$ is 'nearly unbiased' in the sense that

$$0 \leq S(u) - E[\widehat{S}(u)] \leq e^{-\mathcal{N}(u)},$$

where $\mathcal{N}(u) = E[N(u)]$ is the expected number of individuals in the risk set at time u. These comprise the expected number of individuals, either censored or uncensored, which exceed u. Thus, the bias in $\widehat{S}(u)$ is small provided that the expected number at risk is large over the support of the survival distribution; that is, for all values of u.

Because the PL-estimator is built from an interval structure for the observed data, the estimated standard errors may be constructed via Greenwood's Formula in similar circumstances to those described for the lifetable estimate in Chapter 5.

Theorem 6.4 *Let k_u be the index such that $Z'_{(k_u)} < u \leq Z'_{(k_u+1)}$. Then*

$$Var[\widehat{S}(u)] \approx S(u)^2 \sum_{j=1}^{k_u} \frac{q_j}{p_j \mathcal{N}_j},$$

where $\mathcal{N}_j = E[N_j]$. □

The asymptotic normality ensures that the estimation process of reporting $\widehat{S}(u)$ to within two standard errors, where

$$S.E.[\widehat{S}(u)] = \widehat{S}(u) \sqrt{\sum_{j=1}^{k_u} \frac{\widehat{q}_j}{\widehat{p}_j n_j}} \tag{6.3}$$

is reasonable, provided the expected number of censored or uncensored data exceeding u is large for any value of u for which $S(u) > 0$.

Occasionally the approximate 95% confidence interval

$$\widehat{S}(u) \pm 1.96 S.E.[\widehat{S}(u)]$$

may include values outside the parameter space $[0, 1]$. This can occur at specific values of u at the extreme ends of the lifetime distribution. For example, the first line in the software output for the Astrocytomas data of Example 6.1 replaces the upper confidence limit by $u = 1$, the upper bound of the parameter space.

```
         *** Nonparametric Survival ***

time   n.risk   n.event  survival   std.err     95% CI
  0      140       1       0.993    0.00712   0.979 1.000
```

Kalbfleisch and Prentice (1980), among other authors, suggest transforming $Y = \widehat{S}(u)$ using a transformation with unrestricted range. For example,

$$g(Y) = \log_e[-\log_e Y] \quad \text{for} \quad Y = \widehat{S}(u)$$

may be used to create the asymptotic confidence interval for $\log_e[-\log_e S(u)]$ with endpoints

$$g(y) \pm 1.96 S.E.[g(y)].$$

Here, $S.E.[g(y)]$ is easily determined using the Delta Method of Theorem 4.1. Finally, inverting g at the endpoints of this interval creates a new corresponding asymptotic confidence interval back on the original scale entirely in $[0, 1]$. Application of this method to the Astrocytomas data of Example 6.1 at $u = 0.5$ gives an approximate 95% confidence interval for $S(0.5)$ as $(0.950, 0.999)$, compared with the standard $(0.979, 1.007)$ having upper endpoint rounded to 1.00 in the software output.

6.4 Redistribute-to-the-right algorithm

Consider a set of observed right-censored survival data plotted along a time axis, using different notations to differentiate between the censored data points and the uncensored data points. The diagram below uses a standard notation for this representation: hollow dots for right-censored observations; filled-in dots for uncensored observations.

$z_{(1)} \quad z_{(2)} \quad z_{(3)} \; z_{(4)} \qquad\qquad\qquad z_{(n)}$

Recall how such points would be used for plotting the empirical survival function $S_n(u)$ (if we imagine initially that none of the points are censored). $F_n = 1 - S_n$ places probability mass $dF_n = (-dS_n) = \frac{1}{n}$ equally at each of n observed points. This follows by definition of the construction of S_n.

To get an accurate picture of the effect of censoring, we could take the view that the probability mass associated with a right-censored observation should

HAZARD PLOTTING

really be associated with points larger than the observed time of censoring. Following the definition of the empirical survivor function, the probability mass associated with a censor time t_i may be spread equally among all data (both censored and uncensored) which is larger than t_i. The **redistribute-to-the-right algorithm** repeatedly does this for each observed t_i in sequence from smallest to the largest. The redistributions may be set out in tabular form so that the accretions in probability mass in the right tail of the distribution may be monitored. The algorithm provides the values of $d\widehat{F}(z_{(i)}) = [-d\widehat{S}(z_{(i)})]$. The following table shows the redistributions for eight hypothetical data points, written vertically in the table, using the '+' notation to indicate right-censoring.

Observed data	Redist. 1	Redist. 2	Redist. 3
5	$\frac{1}{8}$	$\frac{1}{8}$	$\frac{1}{8}$
13	$\frac{1}{8}$	$\frac{1}{8}$	$\frac{1}{8}$
14	$\frac{1}{8}$	$\frac{1}{8}$	$\frac{1}{8}$
17	$\frac{1}{8}$	$\frac{1}{8}$	$\frac{1}{8}$
18+	$\frac{1}{8}$	0	0
19	$\frac{1}{8}$	$\frac{1}{8} + (\frac{1}{3})(\frac{1}{8})$	$\frac{1}{8} + (\frac{1}{3})(\frac{1}{8})$
20+	$\frac{1}{8}$	$\frac{1}{8} + (\frac{1}{3})(\frac{1}{8})$	0
21+	$\frac{1}{8}$	$\frac{1}{8} + (\frac{1}{3})(\frac{1}{8})$	$\frac{1}{8} + (\frac{1}{3})(\frac{1}{8}) + [\frac{1}{8} + (\frac{1}{3})(\frac{1}{8})]$

After Redistribution 3, the final redistribution needed for this example, the algorithm supplies the values of $d\widehat{F} = (-d\widehat{S})$ at the observed data points as:

$$\frac{1}{8},\ \frac{1}{8},\ \frac{1}{8},\ \frac{1}{8},\ 0,\ \frac{1}{6},\ 0,\ \frac{1}{3}.$$

The redistribute-to-the-right estimator of $S(u)$ is then the sum of all probability masses exceeding u. Notice that the probability mass remains accumulated at the largest observation in accordance with Implementation 3. The proof of the following theorem is developed in three easy stages in Exercises 6.15-6.17.

Theorem 6.5 *The redistribute-to-the-right estimator of S is identical to the PL-estimator of S.* □

6.5 Hazard plotting

The **cumulative hazard function** for a survival random variable Y with hazard function h and survival function S is defined in Definition 4.5 by

$$H(u) = -\log_e S(u).$$

An estimator $\widehat{H}$ of H is defined at each point u by

$$\widehat{H}(u) = -\log_e \widehat{S}(u),$$

where $\widehat{S}$ is the PL-estimator based on observed distinct survival times $z'_{(1)} < z'_{(2)} < z'_{(3)} < \ldots < z'_{(k)}$, $k \leq n$. $\widehat{H}$ is the censored data analogue of the empirical cumulative hazard function discussed in Chapter 4.

The accuracy of $\widehat{H}(u) = -\log_e \widehat{S}(u)$ as an estimate of $H(u) = -\log_e S(u)$ is easily constructed because we know the accuracy of $\widehat{S}(u)$ as an estimate of $S(u)$. Our estimator is a function of an estimator whose properties are known. The variance is easily assessed through the Delta Method of Theorem 4.1. To see this, put $Y = \widehat{S}(u)$. Then $\widehat{H}(u) = -\log_e \widehat{S}(u) = -\log_e Y$ is the choice of function g satisfying $g'(Y) = -\frac{1}{Y}$. Therefore,

$$E[\widehat{H}(u)] \approx -\log_e E[Y] = -\log_e E[\widehat{S}(u)] \approx -\log_e S(u) = H(u)$$

and

$$Var[\widehat{H}(u)] \approx \left(\frac{1}{E(Y)}\right)^2 Var(Y) = \frac{1}{S(u)^2} Var[\widehat{S}(u)] \approx \sum_{j=1}^{k_u} \frac{q_j}{p_j N_j}.$$

Taking square roots gives a standard error which is estimated by again using the PL-estimator. Then, at a fixed point $u = u^*$, the cumulative hazard is estimated by the approximate two standard error confidence interval with endpoints

$$\widehat{H}(u^*) \pm 2 \sqrt{\sum_{j=1}^{k_{u^*}} \frac{\widehat{q}_j}{\widehat{p}_j N_j}}.$$

A first-order approximation, discussed by Nelson (1972) in an early paper on hazard plotting, may be written

$$\widehat{H}(u) \approx \widetilde{H}(u) = \sum_{i: z'_{(i)} \leq u} \frac{d_i}{n_i},$$

using the observed data $z'_{(1)} < z'_{(2)} < z'_{(3)} < \ldots < z'_{(k)}$, $k \leq n$. This simplifying approximation arises through the logarithmic expansion of the right-side of $\widehat{H}(u) = -\log_e \widehat{S}(u)$ with $\widehat{S}(u)$ in product form.

Definition 6.5 *If u is fixed, and the distinct right-censored data are denoted by*

$$(Z'_{(1)}, \delta'_{(1)}), (Z'_{(2)}, \delta'_{(2)}), \ldots, (Z'_{(k)}, \delta'_{(k)}),$$

*and D_j deaths occur at $Z'_{(j)}$, then the **Nelson-Aalen estimator** $\widetilde{H}(u)$ of the cumulative hazard $H(u)$ is defined by*

$$\widetilde{H}(u) = \sum_{i: Z'_{(i)} \leq u} \frac{D_i}{N_i}.$$

□

Aalen (1978) re-established the importance of $\widetilde{H}(u)$ from the point of view of counting processes. In this work he established the estimated standard error as

$$S.E.[\widetilde{H}(u)] = \sqrt{\sum_{i:Z'_{(i)}\leq u} \frac{d_i}{n_i^2}}.$$

There are also interesting parallels here with the simple method of estimating the hazard functions in stem-and-leaf plots using complete data in Equation (1.8) and illustrated in Example 2.1. Essentially, since

$$H(u) = \int_0^u h(y)dy,$$

and (using hat notation to represent estimates) according to the stem-and-leaf plot method of hazard estimation from (1.8),

$$\text{estimate of } h(y) \text{ in } [y, y + \Delta y] = \frac{\widehat{P}(y < Y < y + \Delta y)}{(\Delta y)\widehat{P}(Y > y)},$$

we can write

$$\begin{aligned}
\text{estimate of } H(u) &= \sum_{y\leq u} \widehat{h}(y)\Delta y \\
&= \sum_{y\leq u} \frac{\widehat{P}(y < Y < y + \Delta y)}{(\Delta y)\widehat{P}(Y > y)} \Delta y \\
&= \sum_{y\leq u} \frac{\widehat{P}(y < Y < y + \Delta y)}{\widehat{P}(Y > y)} \\
&= \sum_{i:z'_{(i)}\leq u} \frac{\frac{d_i}{n}}{\frac{n_i}{n}} \\
&= \sum_{i:z'_{(i)}\leq u} \frac{d_i}{n_i} \\
&= \widetilde{H}(u).
\end{aligned}$$

Clearly the quantity $\frac{d_i}{n_i}$, the ratio of observed deaths to the number at risk, is very important in survival and hazard function estimation with censored data. This quantity is a simple estimate of the conditional probability that a subject dies at $z'_{(i)}$ given that they are alive and under observation just prior to $z'_{(i)}$.

6.6 QQ-plots for censored data

QQ-plots for Type II censored data are based on the methods of Chapter 4: if a lifetest is terminated after the observed failures of $r < n$ experimental units, the probability plot is based on the r complete observations using the plotting positions $p_j = \frac{j-\frac{1}{2}}{r}$. The censored points are not plotted.

However, plots for Type I right-censored data require modification. (See, for example, Lawless (1981) and Chambers et al. (1983).) Using the PL-estimator, we construct plotting positions (as defined in Chapter 4)

$$p_j = \frac{\widehat{S}(z'^{-}_{(j)}) + \widehat{S}(z'^{+}_{(j)})}{2} \qquad (6.4)$$

only for values of j representing the points from the ranked data set $z'_{(1)} < z'_{(2)} < \ldots < z'_{(k)}$ at which $\delta_{(j)} = 1$. That is, we begin by ranking the data and constructing the PL-estimator — the plotting positions at the complete uncensored observations (points of discontinuity of $\widehat{S}$) are constructed according to (6.4). The QQ-plot then compares $F^{-1}(p_j)$ with $z'_{(j)}$ at these uncensored points. Note that (6.4) reduces to $p_j = \frac{j-\frac{1}{2}}{n}$ when there is no censoring.

6.7 Exercises

6.1. Determine the PL-estimator at each of the uncensored points for the right-censored lifetime data $(z_i, \delta_i), i = 1, 2, 3, \ldots 10$:

$$(3,1), (3,1), (3,0), (3,0), (4,1), (7,1), (9,1), (9,1), (9,0), (9,0).$$

Sketch a graph of the PL-estimator.

6.2. **Comparing survival prospects for two groups**

Recall from the exercises of Chapter 4, the data from the clinical trial examining steroid-induced remission times (weeks) for leukemia patients.

6-MP		Placebo	
6	17+	1	8
6	19+	1	8
6	20+	2	11
6+	22	2	11
7	23	3	12
9+	25+	4	12
10	32+	4	15
10+	32+	5	17
11+	34+	5	22
13	35+	8	23
16		8	

(*Source*: Lawless, J.F. (1982). *Statistical Models and Methods for Lifetime Data.* Wiley: New York.)

EXERCISES 111

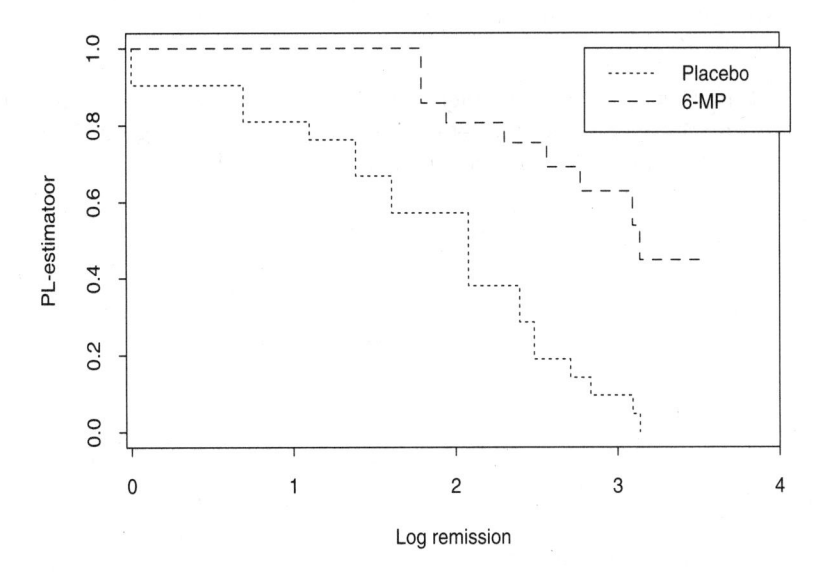

Figure 6.3 *The PL-estimator for the 6-MP and Placebo groups when survival is measured on the* $\log_e$ *scale.*

One group of 21 patients were given 6-mercaptopurine (6-MP); a second group of 21 patients were given a placebo. Since the trial lasted 1 year and patients were admitted to the trial during the year, some of the data could not be gathered by the cut-off date when some patients were still in remission. Right-censored observations Z on remission time Y are given in the table.

As usual, the notation '+' denotes 'right censoring' in the 6-MP group, so that 6+ represents an observed 6-week remission which was still in effect at the closure of the trial.

The S-PLUS graph of the PL-estimator for the 6-MP group is shown in Figure 6.3. On the same axes, the PL-estimator for the placebo group is equivalent to the empirical survivor function since there is no censoring in this group. The plots are shown on the log scale which is often appropriate for the presentation of lifetime data. The PL-estimator was deliberately left 'hanging' in this example to demonstrate what occurs when Implementation 3 is not carried out.

(a) Give a general comparison of the survival prospects of the two groups by interpreting the graphs of the PL-estimators in Figure 6.3.

(b) Read off the graphs estimates of the probability of survival beyond 3 months.

(c) Now use Definition 6.3 to determine the PL-estimator at 3 months for each of the two groups.

6.3. (Continuation) Copy and complete the following table which determines the PL-estimator at the observed remission times for the 6-MP group using the recursive formula (6.1). Missing entries are denoted by □.

$$\begin{aligned}
\widehat{S}(0) &= & 1.000 & & \\
\widehat{S}(6) &= \widehat{S}(0)(\tfrac{18}{21}) &= 0.857 \\
\widehat{S}(7) &= \widehat{S}(6)(\tfrac{16}{17}) &= 0.807 \\
\widehat{S}(10) &= \widehat{S}(7)(\tfrac{14}{15}) &= 0.753 \\
\widehat{S}(13) &= \widehat{S}(10)(\tfrac{11}{12}) &= \square \\
\widehat{S}(16) &= \widehat{S}(13)(\square) &= \square \\
\widehat{S}(22) &= \widehat{S}(16)(\square) &= \square \\
\square &= \widehat{S}(22)(\tfrac{5}{6}) &= \square \\
\widehat{S}(35) &= \square(\tfrac{0}{6}) &= 0.000
\end{aligned}$$

6.4. (Continuation) Show how to use the formula

$$S.E.[\widehat{S}(u)] \approx \widehat{S}(u)\sqrt{\sum_{j=1}^{k_u} \frac{\widehat{q}_j}{\widehat{p}_j N_j}}$$

to give the estimated standard error for $\widehat{S}(10)$ for the 6-MP data.

6.5. In an experiment to gain information on the strength, Y, in coded units, of a certain type of braided cord after weathering, the strengths of 48 pieces of cord that had been weathered for a specified length of time were investigated — these data were extracted from a much larger set in Crowder et al. (1991) for illustrative purposes.

Cord strength

Not damaged

36.3	41.7	43.9	49.9	50.1	50.8	51.9	52.1	52.3	52.3
52.4	52.6	52.7	53.1	53.6	53.6	53.9	53.9	54.1	54.6
54.8	54.8	55.1	55.4	55.9	56.0	56.1	56.5	56.9	57.1
57.1	57.3	57.7	57.8	58.1	58.9	59.0	59.1	59.6	60.4
60.7									

Damaged

26.8	29.6	33.4	35.0	40.0	41.9	42.5

(*Source*: Crowder, M.J., Kimber, A.C., Smith, R.L. and Sweeting, T.J. (1991). *Statistical Analysis of Reliability Data.* Chapman & Hall: London.)

The intention was to obtain the strengths of all 48 pieces of cord. However,

EXERCISES

seven pieces were damaged during the course of the experiment, so that their actual strengths would be greater than the values actually recorded.

(a) Discuss whether these data have been left- or right-censored.

(b) The experimenter thought that, even after weathering, the cord strength should be above 53 in coded units. Determine the Kaplan-Meier estimate $\widehat{S}(53)$ of $S(53)$, where S is the survival function of Y.

Use Greenwood's Formula to compute the standard error of estimate.

6.6. Use computer software to construct the PL-estimator for the Stanford Heart Transplant data of Example 5.2.

Estimate the probability that a Heart Transplant patient will live more than 1 year after transplant. Give the standard error of estimate.

Given that a patient has survived their first 6 months as a Heart Transplant patient, estimate the probability that they will survive the next 6 months.

6.7. The following are the survival times (days) of street lights at various locations around the city: here '+' denotes censoring caused by traffic accidents destroying entire lighting units.

$$36, 37, 38, 38+, 78, 111, 112, 114+, 162, 189, 198, 237, 489+$$

(a) Determine an actuarial estimator of the survival function.

(b) Determine the PL-estimator of the survival function.

6.8. (Continuation) Suppose that we wish to estimate the mean time to failure for street lights in the city.

If Y denotes survival time and Y has survival function S, show that

$$E(Y) = \int_0^\infty S(u)du.$$

Estimate $E(Y)$ by using the PL-estimator and the data of the previous exercise. Estimate the mean life expectancy at age 50.

6.9. (**Jump size in the PL-estimator.**) The obvious generalisation to Theorem 6.1 occurs when there are d_j uncensored ties at $z'_{(j)}$. By following through the proof of Theorem 6.1, show that the size of the discontinuity at $u = z'_{(j)}$ when $\delta'_{(j)} = 1$ is

$$\widehat{S}(z'_{(j-1)}) \frac{d_j}{n_j}$$

when there are d_j uncensored ties at $z_{(j)}$.

6.10. When there is no censoring present in the observed data, explain why $N_1 = N$ and $N_j = N_{j-1} - D_{j-1}$. Using these values, show that $\widehat{S}(u)$ becomes a telescoping product which reverts to the empirical survivor function $S_n(u)$ for estimating $S(u)$.

6.11. Show using Definition 6.3 that the PL-estimator may be determined recursively at adjacent uncensored points, say z'_{j-1} and z'_j, using

$$\widehat{S}(z'_{(j)}) = \widehat{S}(z'_{(j-1)})(1 - \frac{d_j}{n_j}).$$

6.12. Suppose that a component with lifetime Y and survival function S is left to operate until either the component fails and requires replacement, or the component is replaced at a fixed pre-specified time t months under a scheduled replacement policy. This means that the *observed lifetime* of the component is given by the random variable $\min\{Y, t\}$.

The quantity $E[\min\{Y, t\}]$ is called the Expected Service Life of the component.

In order to determine an optimal value of t, data were collected on the the lifetimes Y by observing failure times for the components over many months:

$$1.3, 1.5+, 1.6, 1.8+, 3.2,$$

where '+' denotes right-censoring.

(a) Determine $\widehat{S}(2)$, the product-limit estimate at time 2. Give the standard error of estimate.
(b) Use Efron's redistribute-to-the-right algorithm to estimate $-d\widehat{S}(u)$ at the five data points.
(c) Show that

$$E[\min\{Y, t\}] = \int_0^t S(u)\,du.$$

[Hint: $E[g(Y)] = \int_0^\infty g(u)f(u)\,du$ for a function g.]
(d) Draw a rough sketch of the function $\widehat{S}(u)$ versus u and shade on your graph an area representing an estimate of $E[\min\{Y, 2\}]$
(e) Hence use the product-limit estimator to obtain a numerical estimate for $E[\min\{Y, 2\}]$.
(f) Show how to determine the standard error of estimate in (e) using the Delta Method.

6.13. The cumulative hazard function for a survival random variable Y with hazard function h and survival function S is defined by

$$H(y) = -\log_e S(y).$$

An estimator $\widehat{H}$ of H is defined at each point y by $\widehat{H}(y) = -\log_e \widehat{S}(y)$, where $\widehat{S}$ is the product-limit estimator

$$\widehat{S}(y) = \prod_{i: z'_{(i)} \leq y} \left(\frac{n_i - d_i}{n_i}\right)^{\delta'_{(i)}}$$

EXERCISES

based on observed distinct survival times $z_{(1)} < z_{(2)} < z_{(3)} < \ldots < z_{(k)}$, $k \le n$; $\delta_{(i)}$ is the censor indicator associated with $z_{(i)}$; n_i is the number in the risk set at $z_{(i)}^-$; and d_i is the number who died at $z_{(i)}$.

(a) Show that the size of the discontinuity in $\widehat{S}$ at a point y does depend on $\widehat{S}(y^-)$.

(b) Show that the size of the discontinuity in $\widehat{H}$ at a point y does not depend on $\widehat{S}(y^-)$.

(c) Show that, as a first-order approximation,

$$\widehat{H}(y) \approx \widetilde{H}(y) = \sum_{i:z_{(i)} \le y} \frac{d_i}{n_i}$$

[Hint: $\log_e(1-x) = -x - \frac{1}{2}x^2 - \frac{1}{3}x^3 - \ldots$]

(d) The length of times of remission in acute myelogeneous leukemia under the maintenance chemotherapy for 11 patients are given as

$$9, 13, 13+, 18, 23, 28+, 31, 34, 45+, 48, 161+.$$

Sketch $\widetilde{H}(y)$ as a function of y. Note that $\widetilde{H}(y)$ is piecewise constant with jumps at the uncensored observations.

6.14. Show for Type II censored data (where the first r order statistics $Y_{(1)}, Y_{(2)}, \ldots, Y_{(r)}$ are observed from a total of n observations) that

$$E[H[Y_{(j)}]] = \widetilde{H}[Y_{(j)}],$$

where

$$\widetilde{H}(y) = \sum_{i:y_{(i)} \le y} \frac{d_i}{n_i}.$$

For exponential data, show that a plot of the points $(\alpha_i, y_{(i)})$, $i = 1, 2, \ldots, r$ for

$$\alpha_i = \sum_{j=1}^{i} \frac{1}{n-j+1}$$

should follow a straight line. How should the slope of such a straight line be interpreted?

6.15. Define

$$\widehat{S}(u) = \prod_{i:z_{(i)} \le u} \left(\frac{n-i}{n-i+1}\right)^{\delta_{(i)}}$$

to be the Kaplan-Meier PL-estimator assuming no ties. In this notation, the observed times are $z_i = \min(y_i, t_i)$, for censor constants $t_1, t_2, \ldots, t_n$ and survival times $y_1, y_2, \ldots, y_n$; $\delta_{(i)}$ is the indicator associated with the order statistic $z_{(i)}$.

(a) Show that
$$\widehat{S}(z_{(i)}^-) - \widehat{S}(z_{(i)}) = \prod_{j=1}^{i-1}\left(\frac{n-j}{n-j+1}\right)^{\delta_{(j)}}\left[\frac{\delta_{(i)}}{n-i+1}\right].$$

(b) Show that the bracketed expression [ ] on the right-side of the equation in (a) may be factorised as
$$\frac{\delta_{(i)}}{n}\prod_{j=1}^{i-1}\left(\frac{n-j+1}{n-j}\right).$$

(c) Hence conclude that, when $\delta_{(i)} = 1$, the PL-estimator assigns probability mass
$$\frac{1}{n}\prod_{j=1}^{i-1}\left(\frac{n-j+1}{n-j}\right)^{1-\delta_{(j)}}$$
at $z_{(i)}$.

6.16. With the notation of the previous question, let $j_1 < j_2 < \ldots < j_i$ be the indices of the censored observations which precede $z_{(i)}$. If $\delta_{(i)} = 1$, show that Efron's redistribute-to-the-right algorithm assigns mass
$$\frac{1}{n}\left(1+\frac{1}{n-j_1}\right)\left(1+\frac{1}{n-j_2}\right)\cdots\left(1+\frac{1}{n-j_i}\right)$$
to $z_{(i)}$ (and 0 mass if $\delta_{(i)} = 0$).

6.17. **The redistribute-to-the-right construction is the PL-estimator.** By showing that the formula
$$\frac{1}{n}\prod_{j=1}^{i-1}\left(\frac{n-j+1}{n-j}\right)^{1-\delta_{(j)}}$$
is identical to the formula in the previous exercise, show that the redistribute-to-the-right algorithm gives precisely the PL-estimator.

6.18. When the standard asymptotic confidence interval
$$\widehat{S}(u) \pm 1.96 S.E.[\widehat{S}(u)]$$
for an approximate 95% confidence interval for $S(u)$ at a fixed point u gives values outside the parameter space of the parameter being estimated, Kalbfleisch and Prentice (1980) suggest transforming $\widehat{S}(u)$ using a transformation with unrestricted range.

Let u be a fixed point on the time axis and let $Y = \widehat{S}(u)$.

(a) Write Greenwood's Formula for $S.E.[\widehat{S}(u)]$

(b) Let $g(Y) = \log_e[-\log_e Y]$. Use the Delta Method of Theorem 4.1 to show that
$$Var[g(Y)] \approx \left(\frac{1}{Y\log_e Y}\right)^2 Var[Y].$$

EXERCISES   117

(c) Show that the standard asymptotic approximate 95% confidence interval

$$g(y) \pm 1.96 S.E.[g(y)]$$

for $\log_e[-\log_e S(u)]$ transforms to the standard asymptotic approximate 95% confidence interval

$$\widehat{S}(u)^{\pm 1.96 S.E.[g(y)]}$$

for $S(u)$.

6.19. The following data from Chambers et al. (1983) gives the time to failure, measured in millions of operations, of 40 mechanical devices. Each device had two switches, A or B, which could fail; the failure of either one of these switches was fatal to the operation of the device. Any failure time recorded by the failure of one of the switches, say A, represents a censored observation on the failure time from switch B. In three cases, labelled '-', the device did not fail

| | Failure mode | | | Failure mode | |
Failure time	A	B	Failure time	A	B
1.151		x	2.119		x
1.170		x	2.135	x	
1.248		x	2.197	x	
1.331		x	2.199		x
1.381		x	2.227	x	
1.499	x		2.250		x
1.508		x	2.254	x	
1.534		x	2.261		x
1.577		x	2.349		x
1.584		x	2.369	x	
1.667	x		2.547	x	
1.695	x		2.548	x	
1.710	x		2.738		x
1.955			2.794	x	
1.965	x		2.883	-	-
2.012		x	2.883	-	-
2.051		x	2.910	x	
2.076		x	3.015	x	
2.109	x		3.017	x	
2.116		x	3.793	-	-

(*Source*: Chambers, J.M., Cleveland, W.S., Kleiner, B. and Tukey, P.A. (1983). *Graphical Methods for Data Analysis*. Wadsworth: California.)

(a) Construct the PL-estimators for the failure mode A data and the failure mode B data and superimpose graphs of these on the same diagram for comparison purposes.

(b) Construct censored QQ-plots for each of the two groups, again superimposed on the same diagram, to check whether it is reasonable to assume that the failure times follow lognormal distributions. If this is the case, do these distributions have the same location and scale parameters?

6.20. **An alternative redistribution-of-mass construction.** The following construction of the masses assigned by the PL-estimator to the uncensored points was proposed by Dinse (1985) and provides an alternative to Efron's redistribute-to-the-right algorithm (Efron, 1967).

The Dinse algorithm proceeds as follows:

(a) Rank censored and uncensored observations from smallest to largest and assign equal mass $\frac{1}{n}$ to each observation.

(b) Begin from the far right.

(c) Move to the left until reaching a censored time.

(d) Spread the mass at that censored time (always $\frac{1}{n}$) to all **uncensored** times to its right, in proportion to the masses already accumulated at these times.

(e) Repeat from (c) and stop when the smallest observation is reached.

Assuming no ties, use the formula

$$\frac{1}{n} \prod_{j=1}^{i-1} \left(\frac{n-j+1}{n-j} \right)^{1-\delta_{(j)}}$$

for the probability mass at an uncensored z_i constructed by Efron's method to show that the new Dinse algorithm alters the total mass exceeding a censored time z_j by multiplying all larger uncensored times by

$$\frac{n-j+1}{n-j}.$$

Hence conclude that the Dinse algorithm and Efron's algorithm result in the same estimator.

6.21. Show that (6.4) reduces to $p_j = \frac{j-\frac{1}{2}}{n}$ in the absence of censoring.

CHAPTER 7

Parametric Survival Models under Censoring

In previous chapters we have examined distribution-free estimation of the survival function through lifetables and, more specifically, by using the Kaplan-Meier PL-estimator. Now we adopt the alternative approach of deriving the joint likelihoods for Type I data and for Type II censored data as functions of the parameters in the survival model. The aim is to estimate these parameters by maximising the likelihoods. We will concentrate on the exponential and Weibull survival models.

7.1 Constructing likelihoods

In our usual notation, we seek n observations on a survival random variable Y with survival function S, density f and cumulative distribution function $F = 1 - S$ — all in terms of a parameter of interest θ. For example, when fitting an exponential survival model to lifetime data, θ will be the mean time to failure. In this case the survival function may be written

$$S(y; \beta) = e^{-\frac{y}{\beta}},$$

so that $\theta = E(Y_i) = \beta = \text{MTTF}$.

We will assume that θ is a vector when there is more than one parameter of interest, such as is the case with the Weibull model for which

$$S(y; \theta) = e^{-(\frac{y}{\beta})^{\alpha}},$$

so that here, $\theta = (\alpha, \beta)^T$.

Because of either Type I or Type II censoring, or even truncation, we unfortunately are unable to observe all of the survival data $Y_1, Y_2, Y_3, \ldots, Y_n$. The **maximum likelihood procedures** that we detail in this chapter depend on, as the name suggests, determining an estimator $\widehat{\theta}$ of θ which maximises the likelihood viewed as a function of θ.

Definition 7.1 *The **likelihood**, $L(\theta)$, of the observed data is a constant multiple of the joint distribution of the observed data. The **maximum likelihood estimator** $\widehat{\theta}$ of θ is a function of the observed data which maximises L over values of θ in the parameter space of all possible values of θ.* □

In the case where there is no censoring and we observe the complete random sample $Y_1, Y_2, Y_3, \ldots, Y_n$, the likelihood may be written

$$L(\theta) = c \prod_{i=1}^{n} f(y_i; \theta). \tag{7.1}$$

To simplify notation, we will often suppress the dependence of the survival model on θ and write, for example, $f(y_i)$ rather than $f(y_i; \theta)$. The presence of censoring will make likelihood construction more complex than is given in (7.1). The details for particular censoring schemes are given in the sections which follow. Our discussions in these specific cases will be built from general properties of maximum likelihood estimators.

Since the maximum of a continuous function occurs where the the first derivative is zero (and the second derivative negative), maximum likelihood procedures for the estimation of θ require a solution to the **likelihood equation** $L'(\theta) = 0$, where the derivative is with respect to θ. This locates the value of θ for which $L(\theta)$ is maximum. Usually, the **log-likelihood** $l(\theta) = \log_e L(\theta)$ is easier to maximise, so that the maximum likelihood estimate $\widehat{\theta}$ of θ according to Definition 7.1 satisfies

$$\frac{d}{d\theta} l(\theta)|_{\theta=\widehat{\theta}} = 0.$$

The derivative of the log-likelihood is called the **score function**, $U(\theta)$, so we may write

$$U(\theta) = l'(\theta).$$

Texts such as Cox and Hinkley (1974) establish the theory for the asymptotic properties of maximum likelihood estimators for hypothesis testing and confidence interval construction. We list below the results that are particularly useful to the censored data case. (See also Miller, 1981; Kalbfleisch and Prentice, 1980.) One of the defining characteristics of maximum likelihood theory which shall be used is the **Fisher information**, $I(\theta)$, defined by

$$I(\theta) = E\left[-l''(\theta)\right]. \tag{7.2}$$

The first asymptotic property involving the score function, established through a central limit theorem approach with mild regularity conditions, may be summarised as

$$U(\theta) \sim_a N(0, I(\theta)), \tag{7.3}$$

where $\sim_a$ indicates that an appropriately standardised version of $U(\theta)$ has the asymptotic distribution on the right-side of (7.3).

Turning to maximum likelihood estimation, the solution $\theta = \widehat{\theta}$ to $U(\theta) = l'(\theta) = 0$ is often found through an iterative process. The **Newton-Rhapson method of scoring** is one such process where the score function is expanded in a Taylor series about a fixed value of θ:

$$0 = l'(\widehat{\theta}) = l'(\theta) + (\widehat{\theta} - \theta)l''(\theta) + \ldots$$

CONSTRUCTING LIKELIHOODS

Ignoring derivatives of order higher than two, it follows that

$$\widehat{\theta} \approx \theta + [-l''(\theta)]^{-1}l'(\theta). \tag{7.4}$$

If we replace $-l''(\theta)$ in (7.4) by its expected value $I(\theta)$, and iterate — by determining the right-side for an initial fixed value of θ, obtaining the estimate on the left-side, then feeding it into the right-side as the new value of θ, and so on — the generated sequence of estimates of θ converges to the maximum likelihood estimator $\widehat{\theta}$ under mild regularity conditions. This iterative scheme is therefore generated by

$$\widehat{\theta} \approx \theta + I(\theta)^{-1}U(\theta). \tag{7.5}$$

Finally, because of the asymptotic normal distribution for the score function, $U(\theta)$, as described in (7.3), equation (7.5) gives us insight into the asymptotic normal distribution for $\widehat{\theta}$. In fact, under mild regularity conditions involving the existence and boundedness of the third derivative of the likelihood, the asymptotic distribution may be written

$$\widehat{\theta} \sim_a N\left(\theta, I(\theta)^{-1}\right). \tag{7.6}$$

Therefore, we may estimate the asymptotic variance of the maximum likelihood estimator $\widehat{\theta}$ of θ as

$$Var(\widehat{\theta}) \approx I(\widehat{\theta})^{-1}. \tag{7.7}$$

Hypothesis tests for θ may be based on the asymptotic variance for either the score function or $\widehat{\theta}$. For testing $H_0 : \theta = \theta_0$ against suitable alternatives, the following three test statistics may be used:

1. **Wald test:** $(\widehat{\theta} - \theta_0)^2 I(\widehat{\theta}) \approx \left[\dfrac{\widehat{\theta} - \theta_0}{S.E.(\widehat{\theta})}\right]^2 \sim_a \chi_1^2$;

2. **Likelihood Ratio test:** $-2\log_e\left(\dfrac{L(\theta_0)}{L(\widehat{\theta})}\right) \sim_a \chi_1^2$;

3. **Score test:** $\dfrac{U(\theta_0)^2}{I(\theta_0)} \sim_a \chi_1^2$.

When the expected value in the Fisher information, $I(\widehat{\theta})$, cannot be obtained, the **observed information**, $i(\widehat{\theta})$, for

$$i(\theta) = -l''(\theta)$$

may be used to replace $I(\widehat{\theta})$ in (7.7) and the tests following without affecting the asymptotic distributions. This situation often occurs for right-censored data where censor times t_i may remain undisclosed if the actual lifetime y_i is observed, where $y_i < t_i$.

The generalisation of these test results to more than one parameter is given, for example, in Miller (1981) and results in θ being viewed as a p-dimensional vector rather than a scalar; the Fisher information being a $p \times p$ matrix;

and the asymptotic distributions for the Wald, Likelihood Ratio and Score tests being χ_p^2. We will use the two-parameter generalisation for maximum likelihood estimation with a Weibull model. Here, the inverse of the observed information is a 2×2 matrix holding the asymptotic variances of the maximum likelihood estimates down the main diagonal.

7.2 Likelihood: Type I censoring

In this section we construct the likelihood for fixed right-censored (Type I) data and begin with a review of the notation. Suppose that $Y_1, Y_2, Y_3, \ldots, Y_n$ are right-censored by constants $t_1, t_2, \ldots, t_n$.

The right-censored data that we observe may be written as ordered pairs (Z_i, δ_i), $i = 1, 2, \ldots, n$, where $Z_i = \min\{Y_i, t_i\}$ and

$$\delta_i = \begin{cases} 1 & \text{if } Y_i \leq t_i \quad \text{(uncensored)} \\ 0 & \text{if } Y_i > t_i \quad \text{(censored)}. \end{cases}$$

Assume that $Z_{(1)} \leq Z_{(2)} \leq Z_{(3)} \leq \ldots \leq Z_{(n)}$ with $\delta_{(i)} = \delta_j$ when $Z_{(i)} = Z_j$. That is, assume that the observed data are ranked from smallest to largest and that the censor indicators are always re-ranked according to this order.

Our aim is to determine the joint density of the observed censored data (Z_i, δ_i), $i = 1, 2, \ldots, n$, so that θ may be estimated by the method of maximum likelihood as described in the previous section. This, in turn, provides us with estimates, hypothesis tests and confidence intervals for θ. The following theorem shows how the censoring affects the usual likelihood.

Theorem 7.1 *Under Type I right-censoring with fixed censor times, the joint likelihood, $L(\theta)$, of the observed data (Z_i, δ_i), $i = 1, 2, \ldots, n$ is given by*

$$L(\theta) = c \prod_{i=1}^{n} f(z_i)^{\delta_i} S(z_i)^{1-\delta_i}, \tag{7.8}$$

where c is a constant.

Proof: We begin by determining the joint density $f(z, \delta)$ of Z and δ, where $Z = \min\{Y, t\}$, for indicator δ and t fixed. Clearly, conditionally, when $\delta = 0$ then $Z = t$ with probability 1, so that

$$f(z, 0) = P(Z = t | \delta = 0) P(\delta = 0) = P(\delta = 0) = P(Y > t) = S(t). \tag{7.9}$$

Also, conditionally, when $\delta = 1$, then $Z < t$ and

$$f(z|\delta = 1) = \frac{f(z)}{F(t)}$$

is the conditional density used in the calculation of mean residual lifetime. Again, using the conditional argument,

$$f(z, 1) = f(z|\delta = 1) P(\delta = 1) = [\frac{f(z)}{F(t)}] F(t) = f(z).$$

Combining these representations of $f(z,1)$ and $f(z,0)$ gives

$$f(z,\delta) = \begin{cases} f(z) & \text{if } \delta = 1 \\ S(t) & \text{if } \delta = 0 \end{cases}$$

and this is most succinctly written as

$$f(z,\delta) = f(z)^\delta S(z)^{1-\delta}.$$

The joint likelihood is a constant multiple of the product of $f(z_i, \delta_i)$, $i = 1, 2, 3, \ldots, n$. □

Equation (7.8) may be generalised further to accommodate other types of censoring such as interval-censoring which occurs when the value of observed data are known only up to within an interval. Again, assuming that such an interval-censoring mechanism operates independently of the observed lifetimes, it follows from the first line (7.9) of the proof of Theorem 7.1 that if $\delta = 0$ were to represent an interval-censored observation $[t_1, t_2)$, then the contribution to the likelihood may be determined by

$$f(z,0) = P(Z = t|\delta = 0)P(\delta = 0) = P(\delta = 0) = P(t_1 \leq Y < t_2) = S(t_1) - S(t_2).$$

This means that terms of the form $S(t_1) - S(t_2)$ may be included in the likelihood in (7.8) for interval-censored observations.

This implies a broad generalisation of the likelihood structure of (7.8). We can view this equation as having terms explicitly for observed deaths; right-censoring; left-censoring; interval-censoring — when some or all of these effects occur in the same data set. To construct a likelihood of this general form where censoring mechanisms are independent of lifetimes, we include products of terms:

$f(y)$ for observed death at y;
$S(t)$ for right-censoring at t;
$1 - S(l)$ for left-censoring at l;
$S(t_1) - S(t_2)$ for interval-censoring at $[t_1, t_2)$.

Therefore, (7.8) may be generalised to

$$L(\theta) = c \prod_D f(y) \prod_R S(t) \prod_L [1 - S(l)] \prod_I [S(t_1) - S(t_2)]$$

for D, the set of observed deaths; R, the set of right-censored observations; L, the set of left-censored observations and I, the set of interval-censored observations.

7.3 Likelihood: the effect of truncation

Recall from Definition 5.1 that when observed data are left-truncated, we do not even know of the existence of observations below the truncation point. In this sense, the existence of data that we *can* observe is conditional on it being larger than the truncation point. The contribution to the likelihood by truncated data is therefore conditional.

Under left-truncation, the observed data are ordered pairs (y_i, t_i), $i = 1, 2, 3, \ldots, n$, with $y_i \geq t_i$. Assuming as usual that the times of death are independent of the mechanism generating the truncation constants, we may write the conditional density as

$$f(y|y > t) = \frac{f(y)}{S(t)}.$$

Each ordered pair in the data makes such a contribution to the likelihood so that

$$L(\theta) = c \prod_{i=1}^{n} \frac{f(y_i)}{S(t_i)}.$$

Similar ideas apply for right-truncation. In this case, the observed data are ordered pairs (y_i, t_i), $i = 1, 2, 3, \ldots, n$, with $y_i \leq t_i$. In this case, conditional densities of the form

$$f(y|y < t) = \frac{f(y)}{1 - S(t)}$$

are combined to create the likelihood.

Often, data sets encountered in practice are combinations of complex censoring and truncation mechanisms operating together in the same data set. The above arguments concerning the conditional effect of truncation indicate that where left-truncation occurs in right-censored data, the likelihood is easily obtained by dividing each term $f(z_i)$ and $S(z_i)$ in (7.8) by $S(v_i)$, where v_i is the truncation constant associated with z_i.

7.4 Likelihood: Type II censoring

Only the r smallest lifetimes $Y_{(1)}, Y_{(2)}, \ldots, Y_{(r)}$ in the sample of $n > r$ lifetimes $Y_1, Y_2, Y_3, \ldots, Y_n$ are observed in the Type II censored life test. Therefore, the likelihood for a Type II censored sample is a constant multiple of the joint density of the first r order statistics $Y_{(1)}, Y_{(2)}, \ldots, Y_{(r)}$. Standard methods for determining such a density may be found in more advanced theoretical statistics texts such as Hogg and Craig (1978). Here, we summarise the result as

$$f(y_{(1)}, y_{(2)}, \ldots, y_{(r)}) = \frac{n!}{1!1!\ldots 1!(n-r)!} f(y_{(1)}) f(y_{(2)}) \ldots f(y_{(r)}) S(y_{(r)})^{n-r},$$

so that

$$L(\theta) = c f(y_{(1)}) f(y_{(2)}) \ldots f(y_{(r)}) S(y_{(r)})^{n-r} \qquad (7.10)$$

for a constant c. This likelihood has a similar structure to the Type I censoring likelihood — the complete observations are evaluated at the density; the censored observations are evaluated at the survival function. We find this pattern occurring in the next section as well, where we construct likelihoods for data exposed to random censoring mechanisms.

7.5 Likelihood: random censoring

In a study with a fixed termination date, the point of entry into the study by an individual may be random. This is equivalent to suggesting that the times t_i of censoring are not fixed, but are the outcomes of random variables. This assumption of **random censoring** occurs extensively in the medical literature.

▪ Example 7.1
Random censoring at end-of-study for ovarian cancer patients
Potosky et al. (1993) describe a clinical trial to study epithelial ovarian cancer among Medicare enrollees in the United States. In this study were 3550 Medicare beneficiaries, aged over 65, who were diagnosed with ovarian cancer between 1984 and 1989. The survival times were censored on the patients who were still alive at the end of 1990. The censoring times were measured from date of diagnosis, so they varied considerably between patients as they were brought into the study at various stages during the 6-year period 1984-1989. This is a classic example of random right-censoring. The limited duration of the study was taken as the sole cause of censoring, making it reasonable to assume censoring and lifetimes to be independent. Of course selection processes for patients entering the study, where selection favoured those patients with better survival prospects, would violate this independence between lifetime and censor time. □

Consider the following notation: $Y_1, Y_2, Y_3, \ldots, Y_n$ are right-censored by random censor times $T_1, T_2, T_3, \ldots, T_n$. In this notation, T_i is the random censor time associated with Y_i, for $i = 1, 2, 3, \ldots, n$. The notation for the distributions of these random variables is as follows:

$Y_1, Y_2, Y_3, \ldots, Y_n$ are i.i.d. with survival function S, density f, parameter θ;
$T_1, T_2, T_3, \ldots, T_n$ are i.i.d. with survival function G, density g.

> **Random censorship assumption:** For each i, T_i, the random censor time associated with lifetime Y_i is independent of Y_i. That is, we assume that the survival times are independent of the censor times.

This independence assumption, discussed in Example 7.1, allows us to construct the joint distribution of a censored observation and its corresponding censor time.

Randomly right-censored data has the following structure: we observe the ordered pairs (Z_i, δ_i), $i = 1, 2, \ldots, n$, where $Z_i = \min\{Y_i, T_i\}$ and

$$\delta_i = \begin{cases} 1 & \text{if } Y_i \leq T_i \quad \text{(uncensored)} \\ 0 & \text{if } Y_i > T_i \quad \text{(censored)}. \end{cases}$$

Our aim is to determine the likelihood of the observed censored data of the form (Z_i, δ_i), $i = 1, 2, \ldots, n$, so that θ may be estimated by maximum likelihood. The main result is summarised in Theorem 7.2, which importantly shows the likelihood factorising into two parts, one relating to the censor times and one relating to the lifetimes.

Theorem 7.2 *The joint likelihood, $L(\theta)$, of the observed data (Z_i, δ_i), $i = 1, 2, \ldots, n$ is given by*

$$L(\theta) = c \left[\prod_{i=1}^{n} G(z_i)^{\delta_i} g(z_i)^{1-\delta_i} \right] \left[\prod_{i=1}^{n} f(z_i)^{\delta_i} S(z_i)^{1-\delta_i} \right],$$

where c is a constant.

Proof: As in the fixed censorship case, we begin by determining the joint density of Z and δ, where $Z = \min\{Y, T\}$, for indicator δ and T a random variable independent of Y. Note that

$$\begin{aligned} P(Z < z, \delta = 0) &= P(T < z, Y > T) \\ &= \int_0^z P(T < z, Y > T | T = t) g(t) dt \\ &= \int_0^z S(t) g(t) dt, \end{aligned}$$

and differentiating each side of this expression with respect to z by using the Fundamental Theorem of Calculus for differentiating across an integral sign gives

$$f(z, 0) = S(z) g(z).$$

Repeating this method when the second variable is censored gives

$$\begin{aligned} P(Z < z, \delta = 1) &= P(Y < z, Y \leq T) \\ &= \int_0^z P(Y < z, Y \leq T | Y = y) f(y) dy \\ &= \int_0^z G(y) f(y) dy, \end{aligned}$$

and again differentiating each side of this expression with respect to z:

$$f(z, 1) = G(z) f(z).$$

As in the fixed censor time case, we combine these representations of $f(z, 1)$ and $f(z, 0)$ to write

$$f(z, \delta) = [f(z) G(z)]^{\delta} [g(z) S(z)]^{1-\delta}.$$

The joint likelihood is, by definition, a constant product of $f(z_i, \delta_i)$, $i = 1, 2, 3, \ldots, n$. That is,

$$L(\theta) = c \prod_{i=1}^{n} [f(z_i) G(z_i)]^{\delta_i} [g(z_i) S(z_i)]^{1-\delta_i},$$

which gives the required result on rearrangement. □

The implications of this theorem are important for maximum likelihood estimation involving random censor times. Note that the first bracketed term in L, which we extract as $\left[\prod_{i=1}^{n} G(z_i)^{\delta_i} g(z_i)^{1-\delta_i}\right]$, involves only the distribution of the censor times. If there are no parameters of interest in G and g (that is, G and g do not depend on θ), then this term acts as a constant multiple in L when the maximisation of L is with respect to θ. This returns us to the likelihood of Theorem 7.1 for fixed censor times. Furthermore, if we regard the observed censor times as conditionally fixed at values $t_1, t_2, t_3, \ldots, t_n$, then the bracketed term is factored out of the conditional likelihood, again yielding the fixed likelihood of Theorem 7.1. These arguments are in favour of following a fixed censorship model with the likelihood of Theorem 7.1.

In reality, the assumption that T and Y are independent is unlikely to hold. For example, even in studies such as that described in Example 7.1, the progress results of a study may cause the termination date of the study to be brought forward, violating the independence assumption. Specifically, in the Stanford Heart Transplant study of Example 5.2, doctors tended to admit younger transplant patients into the study after it became clear that age had a significant effect on ultimate survival — these young patients then lasted through the duration of the study to be censored at the termination date — making survival and censoring non-independent.

7.6 The exponential model

In this chapter we have discussed the form of the likelihood function for modelled lifetimes with survival function S and probability density function f under forms of censorship and truncation. In order to go further than just being able to write down the likelihood, indeed to be able to maximise it and estimate parameters in the survival model, we need to look at possible and appropriate models for S. Certainly in engineering applications, the **exponential model** is of prime importance, being one of the most widely applied models to continuous lifetimes. It also has the added utility: the likelihood is easily maximised.

It is important to recognise the exponential distribution as part of a larger family grouping centered around the gamma distribution. In fact, the exponential distribution is a special case of the gamma distribution where the shape parameter is set at 1 and the scale parameter remains positive but unrestricted.

Definition 7.2 *Survival variable Y has a* **gamma distribution** *with parameters α and β and write $Y \sim \gamma(\alpha, \beta)$ if Y has probability density function*

$$f(y) = \frac{1}{\beta^\alpha \Gamma(\alpha)} y^{\alpha-1} e^{-\frac{y}{\beta}}, \quad y > 0, \quad \alpha > 0, \quad \beta > 0,$$

where Γ is the gamma function. □

Unfortunately, the hazard function for the gamma distribution has no simple closed form.

According to the definitions in Section 2.1, since $\Gamma(1) = 1$, it follows that the exponential model is a special case of the gamma model with $\alpha = 1$. That is, $Y \sim \gamma(1, \beta)$ if and only if $Y \sim exp(1)$. Then, if we independently add n of these exponential models together (for example, seeking the time until the nth occurrence of the event of interest in a Poisson process), it can be shown that we obtain a $\gamma(n, \beta)$ distribution. Notice that the first parameter in this distribution $\alpha = n$ is an integer. This type of gamma model is given a special name, namely the **Erlang model**, in engineering applications (Smith, 1998, p201).

A particularly useful member of the gamma family is called the chi-squared distribution. This probability distribution is related to the normal distribution and occupies a prominent role in the area of statistical inference.

Definition 7.3 *In terms of a parametric value, ν, called the* **degrees of freedom**, *which may assume only positive integer values, a random variable Y is said to have a* **chi-squared distribution** *with ν degrees of freedom if $Y \sim \gamma(\frac{\nu}{2}, 2)$. In this case we write $Y \sim \chi^2_\nu$.* □

Lemma 7.1 demonstrates the inter-connectedness of the gamma, exponential and chi-squared distributions through changes of scale. These ideas will be important in our estimation procedures.

Lemma 7.1 *Suppose that the survival variable Y is modelled according to a gamma distribution.*

1. *If $Y \sim \gamma(1, \beta)$, then $Y \sim exp(\beta)$.*
2. *If $Y \sim \gamma(k, 2)$, then $Y \sim \chi^2_{2k}$.*
3. *If $Y \sim \gamma(1, 1)$, then $\beta Y \sim \gamma(1, \beta)$.*
4. *If $Y \sim \gamma(k, 1)$, then $2Y \sim \chi^2_{2k}$.*

Proof: These results are trivially seen by comparing the respective moment generating functions

$$\left(\frac{1}{1-\beta t}\right)^\alpha, \quad \left(\frac{1}{1-2t}\right)^{\frac{\nu}{2}} \quad \text{and} \quad \frac{1}{1-\beta t},$$

of the respective $\gamma(\alpha, \beta)$, χ^2_ν and $exp(\beta)$ models for the various parametric choices. A probability distribution is identified by its moment generating function. □

Maximum likelihood estimation under Type II censoring

Suppose now that $Y_1, Y_2, \ldots, Y_n$ are independent and identically distributed (i.i.d.) $exp(\beta)$ variables subject to Type II censoring. Only $Y_{(1)}, Y_{(2)}, \ldots, Y_{(r)}$

THE EXPONENTIAL MODEL

are observed for a fixed predetermined value of $r \leq n$. Our aim is to estimate β, the MTTF, by maximum likelihood methods. Substitution of the exponential density $f(y) = \frac{1}{\beta}e^{-\frac{y}{\beta}}$ and survival function $S(y) = e^{-\frac{y}{\beta}}$ into (7.10) gives

$$\begin{aligned} L(\beta) &= cf(y_{(1)})f(y_{(2)})\ldots f(y_{(r)})S(y_{(r)})^{n-r} \\ &= [\prod_{i=1}^{r}\frac{1}{\beta}e^{-\frac{y_{(i)}}{\beta}}][e^{-\frac{y_{(r)}}{\beta}}]^{n-r} \\ &= c\frac{1}{\beta^r}\exp[-(\frac{\sum_{i=1}^{r}y_{(i)} + (n-r)y_{(r)}}{\beta})]. \end{aligned}$$

Solving $l'(\beta) = 0$ shows that L is maximised at

$$\widehat{\beta} = \frac{Y}{r}, \qquad (7.11)$$

where

$$Y = \sum_{i=1}^{r} Y_{(i)} + (n-r)Y_{(r)}$$

is called the **total observed lifetime** or the **total time on test**. This is because both the censored and the uncensored lifetimes are added to create Y. The maximum of l is achieved at $\widehat{\beta}$ since $l''(\widehat{\beta}) < 0$. Note that when $r = n$ in (7.11) we obtain the complete data case $\widehat{\beta} = \overline{Y}$, so that the maximum likelihood estimator is the sample mean (as we would expect it to be). Generally, the censored items are included in the numerator of (7.11) but not counted in the denominator.

Confidence intervals, hypothesis tests and other inference procedures follow from the probability distribution for Y described by the following theorem.

Theorem 7.3 *If the lifetime variables $Y_1, Y_2, \ldots, Y_n$ are i.i.d. $exp(\beta)$, then for $Y = \sum_{i=1}^{r} Y_{(i)} + (n-r)Y_{(r)}$, the quantity $\frac{2Y}{\beta}$ has a distribution independent of β:*

$$\frac{2Y}{\beta} \sim \chi^2_{2r}.$$

Proof: The proof outline is as follows. By Theorem 3.6,

$$D_1 = Y_{(1)} \sim exp(\frac{\beta}{n}).$$

Following the proof of Theorem 3.7,

$$D_2 = Y_{(2)} - Y_{(1)} \sim exp(\frac{\beta}{n-1}), \ldots, D_r = Y_{(r)} - Y_{(r-1)} \sim exp(\frac{\beta}{n-r+1}).$$

In terms of these independent exponential differences, the total observed lifetime may be expressed as

$$Y = nD_1 + (n-1)D_2 + (n-2)D_3 + \ldots + (n-r+1)D_r.$$

Recall that the moment generating function of an $exp(\beta)$ random variable is

$\frac{1}{1-\beta t}$. By considering a scale change, the moment generating function of

$$\frac{2(n-k+1)D_k}{\beta}$$

is easily seen to be

$$\frac{1}{1-2t} \quad \text{for each} \quad k = 1, 2, \ldots, r,$$

and does not depend on the index k. Therefore, by independence, the moment generating function of $\frac{2Y}{\beta}$ is

$$\frac{1}{(1-2t)^r},$$

being the product of r identical moment generating functions. By Lemma 7.1 this is the moment generating function of a $\gamma(r, 2)$, or χ^2_{2r} distribution. □

■ Example 7.2
Confidence interval estimate for MTTF

Suppose that we have observed an industrial lifetest to fit an exponential survival model to Type II censored data. Since

$$\frac{2Y}{\beta} \sim \chi^2_{2r},$$

let $\chi^2_{2r,\alpha}$ represents the $(1-\alpha)$th quantile of the χ^2_{2r} distribution, or the upper α point. This means that

$$P(W > \chi^2_{2r,\alpha}) = \alpha \quad \text{when} \quad W \sim \chi^2_{2r}.$$

Then, according to this definition,

$$0.95 = P(\chi^2_{2r,0.975} < \frac{2Y}{\beta} < \chi^2_{2r,0.025}) = P(\frac{2Y}{\chi^2_{2r,0.025}} < \beta < \frac{2Y}{\chi^2_{2r,0.975}})$$

gives the simple representation

$$[\frac{2Y}{\chi^2_{2r,0.025}}, \frac{2Y}{\chi^2_{2r,0.975}}]$$

for a 95% confidence interval for β.

Mann and Fertig (1973) give failure times of aircraft components for a life test in which 10 components were observed until failure when 13 components were placed in a Type II censored life test. The observed failure times (in hours) were

$$0.22, 0.50, 0.88, 1.00, 1.32, 1.33, 1.54, 1.76, 2.50, 3.00.$$

Here, $n = 13, r = 10$ and

$$y = \frac{\sum_{i=1}^{r} y_{(i)} + (n-r)y_{(r)}}{r} = \frac{14.05 + (3)(3.00)}{10} = 2.305,$$

THE EXPONENTIAL MODEL

giving an observed 95% confidence interval
$$\left[\frac{2(23.05)}{34.17}, \frac{2(23.05)}{9.591}\right] = [1.35, 4.80].$$

□

Maximum likelihood under Type I censoring

Again we suppose that $Y_1, Y_2, \ldots, Y_n$ are i.i.d. $exp(\beta)$ variables but now subject to Type I censoring. We observe (Z_i, δ_i), $i = 1, 2, \ldots, n$, where $Z_i = \min\{Y_i, t_i\}$ and

$$\delta_i = \begin{cases} 1 & \text{if } Y_i \leq t_i \quad \text{(uncensored)} \\ 0 & \text{if } Y_i > t_i \quad \text{(censored)}. \end{cases}$$

Our aim is to estimate β, the MTTF, by maximum likelihood methods. Substitution of the exponential density and survival function into Theorem 7.1 gives

$$\begin{aligned} L(\beta) &= c \prod_{i=1}^{n} f(z_i)^{\delta_i} S(t_i)^{1-\delta_i} \\ &= c \frac{1}{\beta^{\sum_{i=1}^{n} \delta_i}} \exp[-(\frac{\sum_{i=1}^{n} z_i}{\beta})] \\ &= c \frac{1}{\beta^R} \exp[-(\frac{Y}{\beta})], \end{aligned}$$

where $R = \sum_{i=1}^{n} \delta_i$ is now a random variable (since we do not know at the outset of the study how many items will be censored) and $Y = \sum_{i=1}^{n} Z_i$ still represents the total observed lifetime.

Solving $l'(\beta) = 0$ shows that L is maximised at

$$\widehat{\beta} = \frac{Y}{R} = \frac{\sum_{i=1}^{n} Z_i}{R}.$$

The probability distribution for $\widehat{\beta}$ is now more difficult to obtain than was the case for Type II censored data since $\widehat{\beta}$ is now a ratio of random variables. The asymptotic likelihood methods of Section 7.1 may be applied. That is,

$$\widehat{\beta} \sim_a N(\beta, I(\beta)^{-1})$$

with the asymptotic standard error estimated by

$$Var(\widehat{\beta}) \approx I(\widehat{\beta})^{-1}.$$

This is an example where we will see, in fact, that the variances need to be estimated by using the observed information rather than the Fisher information. That is, $Var(\widehat{\beta}) \approx i(\widehat{\beta})^{-1}$ because not all the censor times $t_1, t_2, \ldots, t_n$, are known in standard Type I censored data. This example is taken up in the problems at the end of the section where it is shown that

$$i(\widehat{\beta}) = \frac{R}{\widehat{\beta}^2} = \frac{\sum_{i=1}^{n} \delta_i}{\widehat{\beta}^2}.$$

■ Example 7.3
Failure data for electronic flight control packages of the type used on the Poseidon missile system

In reliability life testing of electronic equipment which is not expected to fail, experimental data sets often consist of few observed lifetimes and much censoring due to observation terminating with the equipment still functioning.

Myhre (1983) presents such heavily censored times (in minutes) to failure or 'run-out' (also termed **alive times**) for flight control packages. The paper estimates reliabilities at specific times under the exponential model, and also under a gamma mixture distribution which has a decreasing hazard appropriate for the burn in time of the equipment.

Flight control package failure time (mins)
Observed 1 8 10
Alive 59 72 76 113 117 124 145 149 153 182 320

(*Source*: Myhre, J. M. (1983). A decreasing failure rate, mixed exponential model applied to reliability. *Reliability in the Acquisitions Process.* D.J. DePriest and R.L. Launer (Eds). Marcel Dekker: New York.)

These data are Type I censored with $n = 14$ and $r = \sum_{i=1}^{14} \delta_i = 3$. Using the observed values to determine the maximum likelihood estimate gives

$$\widehat{\beta} = \frac{\sum_{i=1}^{n} z_i}{\sum_{i=1}^{14} \delta_i} = 509.667.$$

The standard error of estimate may be determined by

$$S.E.(\widehat{\beta}) \approx \sqrt{i(\widehat{\beta})^{-1}} = \frac{\widehat{\beta}}{\sqrt{\sum_{i=1}^{n} \delta_i}} = 294.26.$$

The reliability estimates quoted in the paper spring directly from the maximum likelihood estimation: they are estimates

$$S(t, \widehat{\beta}) = e^{-\frac{t}{\widehat{\beta}}}$$

of the exponential survival function $S(t)$ at specific times t. These are as follows:

Reliability estimates at time t

Times	6	10	30	50	100	130
Estimates	0.988	0.980	0.943	0.906	0.821	0.774

Lawless (1982) conducts simulations to examine the adequacy of the maximum likelihood procedures for small sample sizes and reports adequate parameter coverage using confidence intervals with samples as small as 10. □

7.7 The Weibull model

We consider here an introduction to maximum likelihood estimation using data subject to Type I censoring in the context of a Weibull model. A more extensive discussion is given in Lawless (1982).

Given: $Y_1, Y_2, \ldots, Y_n$ are i.i.d. $Weibull(\alpha, \beta)$ variables subject to Type I censoring. By the results of Chapter 2, we may equivalently work with Weibull data on the log scale for, as discussed there, the logarithm of Weibull data gives extreme value data. This means that we will observe (Z_i, δ_i), $i = 1, 2, \ldots, n$, where $Z_i = \min\{\log_e Y_i, \log_e t_i\}$ and δ_i are the usual censor indicators of right-censoring. Note that the transformed variables $X_i = \log_e Y_i$ have extreme value density

$$f(x) = \frac{1}{b} e^{(\frac{x-u}{b})} \exp\left(-e^{(\frac{x-u}{b})}\right),$$

$-\infty < x < \infty, -\infty < u < \infty, b > 0$. Here, u and b are the two parameters of this distribution that require estimation. The survival function is given by

$$S(x) = \exp\left(-e^{(\frac{x-u}{b})}\right).$$

Because functions of maximum likelihood estimators are again maximum likelihood estimators, once $\widehat{u}$ and $\widehat{b}$ have been found, the original Weibull parameters may be estimated through the equations $b = \frac{1}{\alpha}$ and $u = \log_e \beta$ which link the two sets of parameters. That is,

$$\widehat{b} = \frac{1}{\widehat{\alpha}} \quad \text{and} \quad \widehat{u} = \log_e \widehat{\beta}.$$

The maximum likelihood estimates of u and b may be found by substituting the extreme value density and survival function into Theorem 7.1. This gives gives

$$\begin{aligned} L(u,b) &= c \prod_{i=1}^{n} f(z_i)^{\delta_i} S(z_i)^{1-\delta_i} \\ &= c \prod_{i=1}^{n} \left[\frac{1}{b} e^{(\frac{z_i-u}{b})} \exp\left(-e^{(\frac{z_i-u}{b})}\right)\right]^{\delta_i} \left[\exp\left(-e^{(\frac{z_i-u}{b})}\right)\right]^{1-\delta_i}. \end{aligned}$$

On taking logarithms we obtain the log-likelihood (apart from an additive constant)

$$l(u,b) = \log_e L(u,b) = -r\log_e b + \sum_{i=1}^{n} \delta_i(\frac{z_i - u}{b}) - \sum_{i=1}^{n} \exp(\frac{z_i - u}{b}),$$

where $r = \sum_{i=1}^{n} \delta_i$ is the outcome of the random variable representing the number of censored items. The score functions are found by differentiating with respect to each of the parameters:

$$\frac{\partial}{\partial u}l(u,b) = -\frac{r}{b} + \frac{1}{b}\sum_{i=1}^{n} \exp(\frac{z_i - u}{b}) \quad (7.12)$$

$$\frac{\partial}{\partial b}l(u,b) = -\frac{r}{b} - \frac{1}{b}\sum_{i=1}^{n} \delta_i(\frac{z_i - u}{b}) + \frac{1}{b}\sum_{i=1}^{n}(\frac{z_i - u}{b})\exp(\frac{z_i - u}{b}) \quad (7.13)$$

Solving these score functions set equal to 0 provides the maximum likelihood solutions. From (7.12), the equation $\frac{\partial}{\partial u}l(u,b) = 0$ may be re-arranged to give

$$e^{\widehat{u}} = \left(\frac{1}{r}\sum_{i=1}^{n} e^{\frac{z_i}{b}}\right)^{\widehat{b}}.$$

Substitution into $\frac{\partial}{\partial b}l(u,b) = 0$ from (7.13) gives, after simplification,

$$\frac{\sum_{i=1}^{n} z_i \exp(\frac{z_i}{b})}{\sum_{i=1}^{n} \exp(\frac{z_i}{b})} - \widehat{b} - \frac{\sum_{i=1}^{n} z_i \delta_i}{r} = 0.$$

The scale estimator $\widehat{b}$ may be determined from this last equation using numerical techniques: if we set

$$g(b) = \frac{\sum_{i=1}^{n} z_i \exp(\frac{z_i}{b})}{\sum_{i=1}^{n} \exp(\frac{z_i}{b})} - b - \frac{\sum_{i=1}^{n} z_i \delta_i}{r},$$

then $g(b) = 0$ may be solved numerically by putting $b_1 = 1$ and then iterating according to the Newton-Rhapson scheme (where g replaces l' in (7.4))

$$b_{k+1} = b_k - \frac{g(b_k)}{g'(b_k)}, \quad \text{for } k = 1, 2, \ldots$$

until the values of b converge.

The variance of the maximum likelihood estimators may be found by inverting the **observed information matrix** evaluated at the maximum likelihood estimates: If $\widehat{\theta} = (\widehat{u}, \widehat{b})^T$ is the maximum likelihood estimate of $\theta = (u,b)^T$, then

$$Var(\widehat{\theta}) = \begin{pmatrix} Var(\widehat{u}) & Cov(\widehat{u},\widehat{b}) \\ Cov(\widehat{b},\widehat{u}) & Var(\widehat{b}) \end{pmatrix}$$

$$\approx \left(-\frac{\partial^2}{\partial \theta^2}\log_e L(\theta)|_{\theta=\widehat{\theta}}\right)^{-1}$$

THE WEIBULL MODEL

$$= \begin{pmatrix} -\frac{\partial^2 l}{\partial u^2}|_{\theta=\widehat{\theta}} & -\frac{\partial^2 l}{\partial u \partial b}|_{\theta=\widehat{\theta}} \\ -\frac{\partial^2 l}{\partial b \partial u}|_{\theta=\widehat{\theta}} & -\frac{\partial^2 l}{\partial b^2}|_{\theta=\widehat{\theta}} \end{pmatrix}^{-1}$$

■ Example 7.4
Failure data for a mechanical switch

The lifetime data from Nair (1984) and Gulati and Padgett (1996) report lifetest results for a mechanical switch set in constant operation until time of failure, Y, measured by millions of operations. The observed data were heavily right-censored in a pattern as indicated in the table below.

| \multicolumn{6}{c}{Switch failure times} ||||||
(y, δ)		(y, δ)		(y, δ)	
1.151	0	1.965	1	2.349	0
1.170	0	2.012	0	2.369	1
1.248	0	2.051	0	2.547	1
1.331	0	2.076	0	2.548	1
1.381	0	2.109	1	2.738	0
1.499	1	2.116	0	2.794	1
1.508	0	2.119	0	2.883	0
1.534	0	2.135	1	2.883	0
1.577	0	2.197	1	2.910	1
1.584	0	2.199	0	3.015	1
1.667	1	2.227	1	3.017	1
1.695	1	2.250	0	3.793	0
1.710	1	2.254	1		
1.955	0	2.261	0		

(*Source*: Brick, M.J., Michael, J.R. and Morganstein, D. (1989). Using statistical thinking to solve maintenance problems. *Quality Progress*, May.)

These data were analysed in MINITAB as an example of an automated fit of a parametric distribution to right-censored data. The edited output included in this example shows the fit of a Weibull model to the data $U = \log_e Y$. (The log scale is a natural scale of measurement for much survival data.) Many distributional shapes had their fit assessed principally via probability plots; a current version of MINITAB has the capacity to automate the fit of eight potential probability models. The level of software automation means that the actual Newton-Rhapson iterations, as described in this section, do not need to be separately programmed. It is vital, however, to understand the nature of the calculations performed in the software.

A probability plot executed in the software showed some deviation from a sought straight line, but the plotted discrepancies were less severe than for other modelled fits.

```
Censoring Information
Count Uncensored value   17
Right censored value     23
Censoring value: Censor = 0
```

Estimation Method: Maximum Likelihood Distribution: Weibull

Parameter Estimates

Parameter	Estimate	Standard Error	95.0% Normal CI Lower	Upper
Shape	4.0955	0.7441	2.8685	5.847
Scale	1.04120	0.06265	0.92538	1.17152

Log-Likelihood = -9.648

Characteristics of Distribution

	Estimate	Standard Error	95.0% Normal CI Lower	Upper
Mean (MTTF)	0.9450	0.05596	0.8415	1.0613
Standard Deviation	0.2595	0.04542	0.1842	0.3657
Median	0.9521	0.05664	0.8473	1.0698
First Quartile (Q1)	0.7681	0.05696	0.6642	0.8883
Third Quartile (Q3)	1.1276	0.07254	0.9941	1.2792
Interquartile Range	0.3595	0.06614	0.2507	0.5156

Note that the the value 1 does not lie in the confidence interval for the shape parameter, so that the exponential model is unlikely to provide an adequate fit to these data.

□

7.8 Exercises

7.1. The lifetime of a manufactured item is thought to be exponentially distributed. N items are selected at random and put on test for a period of length T. During the test, R items failed at times $y_1 < y_2 < y_3 < \ldots < y_R$, the remaining $N - R$ items being still alive at the end of the test.

Write down the likelihood function for the experiment and derive the maximum likelihood estimator of the mean. State, without proof, the sampling distribution of this estimator and explain how you would use it to obtain a confidence interval for the mean lifetime.

In an experiment with $N = 10$ and $T = 100$ days it was found that seven items failed and three did not. The failure times of the seven failed items were 2, 4, 14, 24, 27, 33 and 51 days, respectively. Obtain a 95% confidence interval for the mean lifetime of an item.

7.2. The lifetime of an article is thought to have an exponential distribution but its mean μ is not known. Twelve such articles were selected at random

EXERCISES

and tested until nine of them failed. The nine observed failure times were:

$$8, 14, 23, 32, 46, 57, 69, 88 \text{ and } 109.$$

(a) Check the conjecture of exponentially distributed lifetimes.
(b) Compute the maximum likelihood estimate of μ.
(c) Obtain a 90% confidence interval for μ.

7.3. Consider the situation for observing right-truncated lifetimes having density f and survival function S. The observed data are ordered pairs (y_i, t_i), $i = 1, 2, 3, \ldots, n$ with $Y_i < t_i$. Write an expression for the joint likelihood of the data assuming, as usual, that the times of death are independent of the mechanism generating the truncation constants.

7.4. When estimating a parameter θ, the likelihood for Type I right-censored data may be expressed as

$$L(\theta) = c \prod_{i=1}^{n} f(z_i)^{\delta_i} S(t_i)^{1-\delta_i}$$

when lifetimes are modelled with probability density function f and survival function S. Show that the log-likelihood l may be expressed, up to an additive constant, in the form

$$l(\theta) = \sum_{i=1}^{n} \delta_i h(z_i) - \sum_{i=1}^{n} H(z_i),$$

where h denotes the hazard function and H denotes the cumulative hazard function.

7.5. Suppose that the notation $\sum^*$ is defined by

$$\sum_{i=1}^{*r} Y_{(i)} = \sum_{i=1}^{r} Y_{(i)} + (n-r)Y_{(r)}$$

when $Y_1, Y_2 \ldots, Y_n$ are Type II censored after r failures.

(a) What does $\sum_{i=1}^{*r} Y_{(i)}$ represent when survival times $Y_1, Y_2, \ldots, Y_n$ are Type II censored after r failures?
(b) Write down the joint density of $Y_{(1)}, Y_{(2)}, \ldots, Y_{(r)}$ in terms of the survival function S and density function f common to $Y_1, Y_2 \ldots, Y_n$.
(c) Show that if $Y_1, Y_2, \ldots, Y_n$ are independent observations of the same extreme value distribution with density

$$f(y) = \frac{1}{b} e^{(\frac{y-u}{b})} \exp\left(-e^{(\frac{y-u}{b})}\right),$$

$-\infty < y < \infty, -\infty < u < \infty, b > 0$, then the maximum likelihood estimators $\hat{u}$ and $\hat{b}$ of u and b satisfy

$$e^{\hat{u}} = \left[\frac{1}{r} \sum_{i=1}^{*r} \exp(\frac{y_{(i)}}{\hat{b}})\right]^{\hat{b}}.$$

when the sample is subjected to Type II censoring after r failures.

7.6. (**Random censoring**). Suppose that the time to death, Y, follows an exponential model with hazard rate λ, and that the random right-censoring time, C, is exponential with hazard rate θ. Let $Z = \min\{Y, C\}$ and

$$\delta = \begin{cases} 1 & \text{if } Y \leq C \quad \text{(uncensored)} \\ 0 & \text{if } Y > C \quad \text{(censored)}. \end{cases}$$

Assume that Y and C are independent.

(a) Find $P(\delta = 1)$.
(b) Determine the probability model for the observed lifetime, Z.
(c) Show that Z and δ are independent.
(d) Let $(Z_1, \delta_1), (Z_2, \delta_2), (Z_3, \delta_3), \ldots, (Z_n, \delta_n)$ be a random sample of observed data. If we fit this model, show that the maximum likelihood estimator $\widehat{\lambda}$ of λ is given by

$$\widehat{\lambda} = \frac{\sum_{i=1}^{n} \delta_i}{\sum_{i=1}^{n} Z_i}.$$

(e) Use (a)–(d) to determine the mean and variance of $\widehat{\lambda}$.

7.7. **Weibull age-replacements**. In an article entitled *Using Statistical Thinking to Solve Maintenance Problems*, Brick et al. (1989) use a Weibull probability plot to detect whether their lifetime data on the 'time to replacement of sinker rolls in a sheet metal galvanising process' may be appropriately modelled by the Weibull distribution. The Weibull probability plot constructed in Example 4.2 showed that a Weibull model was indeed appropriate for these data.

$n = 17$ rollers were observed to have the following lifetimes ranked in order from smallest to largest (measured as a number of 8-hour shifts).

Sinker roll lifetimes (8-hour shifts)
10 12 15 17 18 18 20 20 21
21 23 25 27 29 29 30 35

(*Source*: Brick, M.J., Michael, J.R. and Morganstein, D. (1989). Using statistical thinking to solve maintenance problems. *Quality Progress*, May.)

(a) Fit a two-parameter Weibull model to these data. Give the parameter estimates and their standard errors. Compare these estimates with those obtained from the slope and intercept of the probability plot in Example 4.2.

EXERCISES 139

(b) In order to examine the effect of censoring, re-fit a Weibull model to the observed ranked data $z_{(i)}$ in the above table under the right-censoring pattern

$$1, 1, 0, 1, 1, 1, 0, 1, 1, 0, 0, 1, 1, 1, 1, 0, 1,$$

representing $\delta_{(i)}$, $i = 1, 2, \ldots, 17$, where $\delta_{(i)}$ is the censor time associated with $z_{(i)}$. Give the parameter estimates and their standard errors. Construct a probability plot to examine the goodness-of-fit.

7.8. Babu and Feigelson (1996) note that astronomers often assume that the properties of stars or galaxies follow a power law distribution, which is a special case of the **Pareto distribution**, a model with survival function

$$S(y; \gamma, \alpha) = \begin{cases} (\frac{\gamma}{y})^\alpha & \text{if } y > \gamma \\ 0 & \text{otherwise.} \end{cases}$$

Suppose that $(y_1, t_1), (y_2, t_2), (y_3, t_3), \ldots, (y_n, t_n)$ is a left-truncated data set, where t_i is the truncation constant associated with y_i. Estimate the parameters α and γ by the method of maximum likelihood.

7.9. A particular electrical device has a lifetime distribution adequately modelled by an $exp(\theta)$ distribution with mean θ. In setting up a screening procedure for consignments of these devices, it is decided to institute a *Type II censored lifetest plan*:

Accept $H_0 : \theta = 1000$ with probability 0.9 when $\theta = 1000$ hours and probability 0.05 when $\theta = 300$ hours.

(a) Show that, in general, if α and β represent the Type I and Type II errors and r denotes the smallest number of lifetimes which must be observed, then

$$\frac{\chi^2_{2r,1-\alpha}}{\chi^2_{2r,\beta}} = 0.3.$$

(b) Determine the value of r so that the error rate specifications hold.

(c) Determine the value of the constant C so that the decision rule for the test is

$$reject\ H_o\ if\ \widehat{\theta} \leq C,$$

where $\widehat{\theta}$ is the MLE.

(d) Show that the expected duration of the plan is

$$\theta \sum_{i=1}^{r} \frac{1}{n-i+1},$$

where r is chosen as in (b).

7.10.(a) Suppose that a survival random variable Y has an exponential distribution with mean $\frac{1}{\rho}$. When Y is subjected to censoring at time t, then we observe $Z = \min\{Y, t\}$.

Show that π, the probability of censoring, is given by $\pi = e^{-\rho t}$.

Hence show that $\rho E(Z) = 1 - \pi$.

(b) Suppose that W is an absolutely continuous survival random variable with cumulative distribution function F and cumulative hazard function $H(u) = -\log_e[1 - F(u)]$.

Show that if we set $Y = H(W)$, then Y has an exponential distribution with $\rho = 1$.

Hence show that if c is a constant such that $Q = \min\{W, c\}$, then
$$H(Q) = \min\{H(W), H)(c)\}$$
and
$$E[H(Q)] = F(c).$$

7.11. Let h_0 and H_0 be known 'baseline' hazard and cumulative hazard functions, respectively. A survival distribution with parameter ρ, density $f(u; \rho)$, survival function $S(u; \rho)$ and hazard $h(u; \rho)$ follows the **proportional hazards model** if
$$h(u; \rho) = \rho h_0(u).$$

The aim is to estimate the parameter ρ.

(a) Show that the log-likelihood of the observed Type I right-censored sample $Z_1, Z_2, \ldots, Z_n$ (with t_i and δ_i the censor time and censor indicator associated with Z_i) is given by
$$l(\rho) = n_u \log_e \rho + \sum_{i=1}^{n} \delta_i \log_e h_0(z_i) - \sum_{i=1}^{n} \rho H_0(z_i),$$
where n_u denotes the number of uncensored observations.

(b) Hence show that the maximum likelihood estimator of ρ is the **standardised mortality ratio**
$$\hat{\rho} = \frac{n_u}{\sum_{i=1}^{n} H_0(z_i)}.$$

(c) Determine the observed information for this problem; hence give an approximate 95% confidence interval for ρ.

7.12. Suppose that n components have independent lifetimes $Y_1, Y_2, \ldots, Y_n$. Suppose that S_i, the survival function for the ith component, is given by
$$S_i(y) = e^{-(\frac{y}{\beta})^2}.$$

(a) If the components are linked in series, determine the survival function and hazard function for the system.

(b) Give the probability distribution for the lifetime of the series system in (a). Identify the parameters of the distribution.

EXERCISES

(c) Suppose now that the independent lifetimes $Y_1, Y_2, \ldots, Y_n$ are right-censored by corresponding censor times $t_1, t_2, \ldots, t_n$ and that the censoring process $Z_i = \min(Y_i, t_i)$ is recorded by corresponding censor indicators $\delta_1, \delta_2, \ldots, \delta_n$.

Show that the log-likelihood based on the censored data is given by

$$\log l = c - 2r \log \beta - \sum_{i=1}^{n} \frac{z_i^2}{2r},$$

where $r = \sum_{i=1}^{n} \delta_i$ and c is a constant.

Hence determine the maximum likelihood estimator of β.

7.13. Let $Y_1, Y_2, \ldots, Y_n$ denote the modelled $exp(\lambda)$ lifetimes (with mean λ) of n individuals. For each $i = 1, 2, \ldots, n$, we observe (Z_i, δ_i) where

$$Z_i = \min\{Y_i, t_i\}$$

for fixed censor constants t_i.

(a) State the value of $E(Z_i \mid \delta_i = 0)$ and show that

$$E(Z_i \mid \delta_i = 1) = \lambda - \frac{t_i(1 - \rho_i)}{\rho_i},$$

where $\rho_i = 1 - exp(-t_i/\lambda)$. Hence determine $E(Z_i)$.

(b) Show that

$$I(\lambda) = E\left(-\frac{d^2 \log L}{d\lambda^2}\right),$$

the Fisher information in the sample, is given by

$$I(\lambda) = \frac{\sum_{i=1}^{n} \rho_i}{\lambda^2}.$$

(c) If $\widehat{\lambda}$ denotes the maximum likelihood estimator of λ, then $I(\widehat{\lambda})$ (which is used in statistical tests and confidence intervals) is easily seen to involve all of $t_1, t_2, \ldots, t_n$, some of which may remain undisclosed. Instead we utilise the equivalent asymptotic MLE methods based on the *observed information*

$$i(\widehat{\lambda}) = -\frac{d^2 \log L}{d\lambda^2} \Big|_{\lambda = \widehat{\lambda}}.$$

Show that

$$i(\widehat{\lambda}) = \frac{\sum_{i=1}^{n} \delta_i}{\widehat{\lambda}^2}.$$

7.14. Let $Y_1, Y_2, \ldots, Y_n$ be independent and identically distributed random variables with **extreme value** probability density function

$$f(y; \theta) = e^{-(y-\theta)} \exp[-e^{-(y-\theta)}],$$

where $-\infty < \theta < \infty$.

(a) Find $E(Y_i)$ for any $i = 1, 2, 3, \ldots, n$.
(b) Find a maximum likelihood estimator of θ.
(c) Let Γ denote the gamma function of Definition 2.3. Show that

$$\widehat{\theta} = \frac{\Gamma^{(1)}(n)}{\Gamma(n)} - \log\left(\sum_{i=1}^{n} e^{-Y_i}\right)$$

is an unbiased estimator of θ in the sense that $E(\widehat{\theta}) = \theta$.

CHAPTER 8

Fitting Parametric Regression Models

In this chapter we examine the relationship between the survival time Y and values of an **explanatory variable** X. For example, if Y denotes 'survival time following heart transplant', X may denote 'age'; the aim is to use X in modelling for Y. In this general framework for the regression of Y on X, other terms sometimes used to describe X are **covariate** and **regressor variable**. We will use such terms interchangeably.

More generally, we may expect survival time to depend on the outcome of several explanatory variables; these may be collected together in vector form, $\mathbf{X}$. In the heart transplant case, we may consider $\mathbf{X}^T = (X_1, X_2, X_3) = $ (age, sex, treatment type). In general, $\mathbf{X}$ is a vector of p explanatory variables and for a particular subject, individual or case, we may observe a vector $\mathbf{X} = \mathbf{x}$, where

$$\mathbf{x}^T = (x_1, x_2, \ldots, x_p).$$

The notation $Y_\mathbf{x}$ is used to represent response Y for covariate values $\mathbf{X} = \mathbf{x}$.

Much of the fundamental discussion of this section is built from results in Lawless (1982), but current trends require facility at fitting models, such as parametric regression models, using statistical software. This requires not only a heightened awareness of the fundamentals, but also quick checks for model suitability, diagnostics and interpretation of results. Primarily, what is needed is a deep understanding of how various models operate.

We present two types of modelling processes — 'proportional hazards' and 'accelerated lifetimes'. We shall see that only in the case of the Weibull models do these two kinds of modelling processes coincide. If it is not appropriate to fit a Weibull model to given data, then a choice needs to be made between using accelerated lifetimes or proportional hazards.

8.1 Basic concepts of proportional hazards

We begin with parametric descriptions of possible models relating X and Y. Since the response will be modelled to depend on the observed value of a vector of explanatory variables $\mathbf{x}$, clearly the hazard function will also depend on $\mathbf{x}$. Suppose that $h_\mathbf{x}(y)$ denotes the hazard function of Y when $\mathbf{x}$ is the vector of observed covariate values. The question then arises: what effect do the covariates have on the hazard function and how should this effect be modelled? One of the most commonly used models suggests that the covariates have a multiplicative effect on a basic hazard function called the **baseline** hazard. This is the stuff of the so-called 'proportional hazards model' which Definition 8.1 makes precise.

Definition 8.1 *Let $Y_{\mathbf{x}}$ denote the response depending on an observed vector of covariate values $\mathbf{x}$. By a* **proportional hazards model** *for $Y_{\mathbf{x}}$, we mean the model*

$$h_{\mathbf{x}}(y) = h_0(y)g_1(\mathbf{x}),$$

where g is a positive function of $\mathbf{x}$ and $h_0(y)$ is called the **baseline hazard**, *representing the hazard function for an individual having $g_1(\mathbf{x}) = 1$.* □

It is clear that according to this definition, the covariates have a multiplicative effect on the hazard through the product of h_0, which does not depend on $\mathbf{x}$, and g, which does depend on $\mathbf{x}$. The 'proportional' terminology arises in a perfectly natural way: if two experimental subjects have lifetimes depending on respective vectors of covariate values $\mathbf{x}_1$ and $\mathbf{x}_2$, then

$$\frac{h_{\mathbf{x}_1}(y)}{h_{\mathbf{x}_2}(y)} = \frac{h_0(y)g_1(\mathbf{x}_1)}{h_0(y)g_1(\mathbf{x}_2)} = \frac{g_1(\mathbf{x}_1)}{g_1(\mathbf{x}_2)},$$

showing clearly how the baseline hazards cancel from this ratio, so that the hazard ratio for the two experimental subjects does not depend on y.

The function g_1 in Definition 8.1 is required to satisfy $g_1(\mathbf{x}) \geq 0$ and $g_1(\mathbf{0}) = 1$ only. This is often accomplished by introducing a vector $\boldsymbol{\beta}$ of p parameters by setting $\mathbf{x}^T = (x_1, x_2, \ldots, x_p)$, $\boldsymbol{\beta}^T = (\beta_1, \beta_2, \ldots, \beta_p)$ and

$$g_1(\mathbf{x}) = e^{\boldsymbol{\beta}^T \mathbf{x}}. \tag{8.1}$$

This works because when $\mathbf{x} = \mathbf{0}$ in (8.1), $g_1(\mathbf{0}) = e^{\boldsymbol{\beta}^T \mathbf{0}} = e^0 = 1$ and the exponential function is always positive. Notice that in (8.1), $\boldsymbol{\beta}^T \mathbf{x} = \beta_1 x_1 + \beta_2 x_2 + \ldots + \beta_p x_p$ is a linear combination of the covariate values from the vector $\mathbf{x}$, where the coefficients are taken from the vector of parameters $\boldsymbol{\beta}$. The standard proportional hazards model in Definition 8.1 then becomes

$$h_{\mathbf{x}}(y) = h_0(y) e^{\boldsymbol{\beta}^T \mathbf{x}}. \tag{8.2}$$

It follows that the baseline hazard occurs when $\mathbf{x} = \mathbf{0}$, the zero vector, in which case $h_{\mathbf{0}}(y) = h_0(y)$ for all y. When the representation in (8.2) is used, the process of actually *fitting* a proportional hazards model will involve the estimation of the p parameters in $\boldsymbol{\beta}$ using the observed responses and covariate values. This will be discussed in Section 8.6 for a parametric approach and in the next chapter for a distribution-free approach.

We have seen the implication of the proportional hazards assumption on hazards. What, then, are the implications of the proportional hazards assumption on the survival function? If we define $S_{\mathbf{x}}(y)$ to be the survival function for an individual with hazard function $h_{\mathbf{x}}(y)$, then it follows from the sur-

BASIC CONCEPTS OF PROPORTIONAL HAZARDS 145

vival/hazard relationships of Chapter 1 that

$$\begin{aligned} S_{\mathbf{x}}(y) &= \exp(-\int_0^y h_{\mathbf{x}}(u)du) \\ &= \exp(-\int_0^y h_0(u)g_1(\mathbf{x})du) \\ &= \exp(-g_1(\mathbf{x})\int_0^y h_0(u)du) \\ &= \left[\exp(-\int_0^y h_0(u)du)\right]^{g_1(\mathbf{x})} \\ &= [S_0(y)]^{g_1(\mathbf{x})}, \end{aligned}$$

where

$$S_0(y) = \exp(-\int_0^y h_0(u)du)$$

is the baseline survival function $S_0(y) = S_{\mathbf{0}}(y)$, which occurs when $\mathbf{x} = \mathbf{0}$. In summary, the effect of the proportional hazards assumption on $S_{\mathbf{x}}(y)$ is that it specifies a positive function g_1 and a baseline survival function S_0 such that

$$S_{\mathbf{x}}(y) = [S_0(y)]^{g_1(\mathbf{x})}. \tag{8.3}$$

This shows that in a proportional hazards model, the vector of covariates provides a 'power' effect on the baseline survival function.

The following example demonstrates this effect when there is a single covariate in the model ($p = 1$). This covariate is used to denote group membership of either Group 0 or Group 1 — convenient labels perhaps denoting 'control' and 'treatment' groups in a clinical trial.

■ Example 8.1
Proportional hazards with a group membership covariate
Suppose that two groups of patients (Group 0 and Group 1) are identified by the covariate X = group membership, where x_i the observed value X for Patient i is given by

$$x_i = \begin{cases} 0 & \text{if Patient } i \text{ is from Group 0} \\ 1 & \text{if Patient } i \text{ is from Group 1.} \end{cases}$$

In this example there are $p = 1$ covariates, so that the vector $\mathbf{x}$ is really a scalar x. Let $S_0(y)$ denote the survival function for the patients in Group 0. If x is the covariate value for a patient from Group 1,

$$g_1(\mathbf{x}) = e^{\boldsymbol{\beta}^T \mathbf{x}} = e^{\beta x} = e^{\beta} = \gamma > 0.$$

That is, $g_1(1) = \gamma$. Similarly, if x is the covariate value for a patient from Group 0,

$$g_1(x) = e^{\beta x} = e^0 = 1,$$

so that $g_1(0) = 1$. In terms of survival functions, $S_{\mathbf{x}}(y) = [S_0(y)]^{g_1(\mathbf{x})}$ reduces, for $x = 1$, to

$$S_1(y) = [S_0(y)]^{g_1(1)} = S_0(y)^{\gamma}$$

and this represents the survival function for patients in Group 1. In summary, S_i, the survival function for patients in Group i, satisfies

$$S_1(y) = S_0(y)^{\gamma}$$

and S_0 and S_1 comprise what is termed a family of **Lehmann alternatives** in that their survival functions are powers of one another. □

■ Example 8.2
Hazard functions for individuals

Collett (1994) discusses the possibility of different experimental subjects having specific hazard functions related to the baseline hazard. When we examine the lifetimes of n experimental units (or patients in a medical context), Unit 1, Unit 2, ..., Unit n, where Unit i has a set of p covariates arranged in vector form with observed value $\mathbf{x} = \mathbf{x}_i = (x_{i1}, x_{i2}, \ldots, x_{ip})^T$, it is convenient to model the lifetime Y_i for Unit i with survival function S_i, density f_i, hazard h_i by

$$h_i(y) = h_{\mathbf{x}_i}(y) = h_0(y)g_1(\mathbf{x}_i), \qquad S_i(y) = S_{\mathbf{x}_i}(y), \qquad f_i(y) = f_{\mathbf{x}_i}(y).$$

In this notation, x_{ik} is the kth covariate value of the ith unit for $1 \leq i \leq n$ and $1 \leq k \leq p$. The proportional hazards model then ensures that

$$\frac{h_i(y)}{h_0(y)} = e^{\boldsymbol{\beta}^T \mathbf{x}_i}.$$

Again, note that $\mathbf{x}_i$ is not the ith covariate: it is the observed vector of different covariate values measured on the ith unit. The structure of this ratio of hazards ensures that the hazard function for each individual unit or patient is a different multiple of the baseline hazard. The whole point of the modelling process is that this multiple may be estimated simply through the estimation of $\boldsymbol{\beta}$. □

8.2 Proportional hazards for Weibull data

Recall from Chapter 2 that the survival function of $Y \sim Weibull(\alpha, \beta)$ is given by

$$S(y) = \exp[-(\frac{y}{\beta})^{\alpha}], \quad \text{for } y > 0.$$

Covariates are usually introduced into the Weibull model in the position of the scale parameter β. That is, we allow the scale parameter to become some function $w(\mathbf{x})$ of the covariates and set

$$S_{\mathbf{x}}(y) = \exp[-(\frac{y}{w(\mathbf{x})})^{\alpha}]. \tag{8.4}$$

BASIC CONCEPTS OF ACCELERATED LIFETIMES   147

We wish to show that such an arrangement for including covariates is compatible with the proportional hazards structure of Definition 8.1. To do this we will want to use
$$g_1(\mathbf{x}) = e^{\boldsymbol{\beta}^T \mathbf{x}}$$
as before, expressed in terms of the vector of parameters $\boldsymbol{\beta}$ and vector, $\mathbf{x}$, of covariates. Suppose that we specifically choose $w(\mathbf{x})$ in the following way:
$$w(\mathbf{x}) = g_1(\mathbf{x})^{-\frac{1}{\alpha}}.$$
What hazard function can we derive from this choice of w? From the structure of the Weibull hazard function in (2.2), it follows that $S_{\mathbf{x}}$ satisfies the conditions of the proportional hazards model in that
$$h_{\mathbf{x}}(y) = \alpha y^{\alpha-1} w(\mathbf{x})^{-\alpha} = h_0(y) g_1(\mathbf{x}),$$
so that Definition 8.1 is satisfied for $h_0(y) = \alpha y^{\alpha-1}$, which is the hazard function of a $Weibull(\alpha, 1)$ survival variable which does not depend on $\mathbf{x}$.

This means that when we outfit a Weibull model
$$Y_{\mathbf{x}} \sim Weibull(\alpha, w(\mathbf{x}))$$
with proportional hazards, the **baseline hazard** is
$$h_0(y) = \alpha y^{\alpha-1},$$
and the hazard for an individual with covariates in $\mathbf{x}$ is
$$h_{\mathbf{x}}(y) = h_0(y) e^{\boldsymbol{\beta}^T \mathbf{x}}.$$
Software allows us to estimate α and $\boldsymbol{\beta}$ by the method of maximum likelihood.

8.3 Basic concepts of accelerated lifetimes

In our standard notation, suppose that $S_{\mathbf{x}}(y)$ denotes the survival function of $Y_{\mathbf{x}}$, the observed response for the vector of covariates $\mathbf{x}$. In an **accelerated failure time model**, the covariates are assumed to act directly on lifetime, so as to speed it up or retard its progress. In terms of the response variable, the speeding up or slowing down is accomplished by a positive covariate function g_2 and we may write
$$Y_{\mathbf{x}} g_2(\mathbf{x}) = Y_0, \qquad (8.5)$$
where $g_2(\mathbf{0}) = 1$. When $\mathbf{x} = \mathbf{x}_i$, and we observe the particular covariate values associated with the ith response, we use the usual response variable notation Y_i. That is, from (8.5),
$$Y_i = Y_{\mathbf{x}_i} = Y_0 [\frac{1}{g_2(\mathbf{x}_i)}].$$
How does this affect the survival function for $Y_{\mathbf{x}}$? To see this we simply calculate the probability that $Y_{\mathbf{x}}$ exceeds y. Using (8.5),
$$S_{\mathbf{x}}(y) = P(Y_{\mathbf{x}} > y) = P(\frac{Y_0}{g_2(\mathbf{x})} > y) = P(Y_0 > y g_2(\mathbf{x})) = S_0[y g_2(\mathbf{x})]. \quad (8.6)$$

This equation becomes the characterising feature of the accelerated failure time model.

Definition 8.2 *By an* **accelerated failure time model** *for $Y_\mathbf{x}$, we mean the model*
$$S_\mathbf{x}(y) = S_0[y g_2(\mathbf{x})],$$
where g_2 is a positive function of the covariate $\mathbf{x}$ and $S_0(y)$ is called the **baseline survival function**, *representing the survival function for an individual for which $g_2(\mathbf{x}) = 1$.* □

The usual choice of g_2 is $g_2(\mathbf{x}) = e^{-\boldsymbol{\beta}^T \mathbf{x}}$, since for this definition of g_2 we have $g_2 \geq 0$ and $g_2(\mathbf{0}) = 1$. This means that
$$Y_\mathbf{x} = Y_0 e^{\boldsymbol{\beta}^T \mathbf{x}},$$
and if $\boldsymbol{\beta}^T \mathbf{x} < 0$, then the covariate has caused the lifetime to be shortened from Y_0 to $Y_\mathbf{x}$; similarly, if $\boldsymbol{\beta}^T \mathbf{x} > 0$, then the covariate has caused the lifetime to be lengthened (accelerated) from Y_0 to $Y_\mathbf{x}$.

Taking natural logarithms on each side of (8.5) gives
$$\begin{aligned} \log_e Y_\mathbf{x} &= \log_e Y_0 - \log_e g_2(\mathbf{x}) \\ &= \log_e Y_0 + \boldsymbol{\beta}^T \mathbf{x} \\ &= \mu_0 + \boldsymbol{\beta}^T \mathbf{x} + \sigma Z, \end{aligned}$$
where $\log_e Y_0 = \mu_0 + \sigma Z$ and Z is a random variable with mean 0 and variance 1. This provides us with the **log-linear model**
$$\log_e Y_\mathbf{x} = \mu(\mathbf{x}) + \sigma Z, \tag{8.7}$$
where $\mu(\mathbf{x}) = \mu_0 + \boldsymbol{\beta}^T \mathbf{x}$.

Finally, notice that by differentiating the survival function $S_\mathbf{x}$ in Definition 8.2 we find that the accelerated lifetime model has associated hazard function $h_\mathbf{x}(y) = h_0[y g_2(\mathbf{x})] g_2(\mathbf{x})$.

8.4 Accelerated lifetimes for Weibull data

The discussion of the previous section indicates that in an accelerated lifetime model, the logarithm of lifetime is a linear function of $\mathbf{x}$. This naturally occurs when lifetimes follow a Weibull distribution. The results from Section 2.3 concerning the Weibull distribution show that if we model $Y_\mathbf{x} \sim Weibull(\alpha, w(\mathbf{x}))$, then $Y_\mathbf{x}$ has survival function
$$S_\mathbf{x}(y) = \exp[-(\frac{y}{w(\mathbf{x})})^\alpha].$$

We now want to show that by careful choice of $w(\mathbf{x})$ we arrive at the log-linear model of (8.7). For this construction, set
$$w(\mathbf{x}) = g_2(\mathbf{x})^{-1} = e^{\boldsymbol{\beta}^T \mathbf{x}}.$$

It follows from the extreme value survival function derivation in Section 2.3 that if $Y \sim Weibull(\alpha, \beta)$, then

$$\log_e Y \sim extremevalue(u, b), \quad u = \log_e \beta \text{ and } b = \frac{1}{\alpha}$$

and, through standardising,

$$\frac{\log_e Y - u}{b} \sim extremevalue(0, 1).$$

Applying this to the responses $Y_\mathbf{x}$, it follows that

$$Z = \frac{\log_e Y_\mathbf{x} - u(\mathbf{x})}{b} \sim extremevalue(0, 1) \qquad (8.8)$$

for $u(\mathbf{x}) = \log_e w(\mathbf{x}) = \boldsymbol{\beta}^T\mathbf{x}$ and $b = \frac{1}{\alpha}$. Thus, (8.7) and (8.8) represent essentially the same model when Z has an extreme value distribution with parameters 0 and 1 and the scale parameter is $\sigma = b = \frac{1}{\alpha}$. Therefore, when the log-linear model of (8.7) is applied to Weibull data, we have

$$\log_e Y_\mathbf{x} = \mu_0 + \boldsymbol{\beta}^T\mathbf{x} + \sigma Z, \quad \text{for } Z \sim extremevalue(0, 1). \qquad (8.9)$$

This parallels the case of simple loglinear models for lognormal survival times in which we expect residuals to be normally distributed with mean 0 and variance σ^2.

Accelerated lifetime models and proportional hazards models are, in essence, very different in their requirements for a good fit to data. In the next section we give some diagnostic plots to assist in choosing between the two models for a given data set. Now we note that in the case of Weibull data, the two models coincide.

Theorem 8.1 *The proportional hazards model and the accelerated lifetime model coincide if and only if the lifetimes follow a Weibull distribution.*

Proof: For covariate functions g_1 and g_2, we must show that $S_\mathbf{x}(y)$, the survival function of $Y_\mathbf{x}$, satisfies $S_\mathbf{x}(y) = S_0[yg_2(\mathbf{x})] = [S_0(y)]^{g_1(\mathbf{x})}$ if and only if there exist $\alpha > 0$ and $\beta > 0$ such that $S_0(y) = e^{-(\frac{y}{\beta})^\alpha}$. The 'if' part of the proof has been discussed already in the construction of (8.3) and (8.6). For the 'only if' proof, we supply an outline and leave the main details to the exercises.

Suppose that $S_0[yg_2(\mathbf{x})] = [S_0(y)]^{g_1(\mathbf{x})}$. We need to show that S_0 is a Weibull survival function. Taking logarithms twice in this equations yields

$$\log_e g_1(\mathbf{x}) + \log_e[-\log_e S_0(y)] = \log_e\{-\log_e S_0[yg_2(\mathbf{x})]\}.$$

Re-labeling the lifetime axis with $y = e^t$ and setting $G_0(t) = \log_e[-\log_e S_0(y)]$ gives a re-expression of the previous equation as:

$$\log_e g_1(\mathbf{x}) + G_0(t) = G_0[t + g_3(\mathbf{x})],$$

where $g_3(\mathbf{x}) = \log_e g_2(\mathbf{x})$. This ensures that G_0 is a linear function $G_0(t) = a + kt$ and from this the required form of S_0 follows. □

8.5 Diagnostics for choosing between models

Lawless (1982) suggests many diagnostic tests that may be used for checking the appropriateness of both the accelerated lifetime model and the proportional hazards model. In this section we concentrate on some of these suggestions, many of which exploit the simple plotting methods of Chapter 4.

We have seen that the accelerated lifetime model may be written in log-linear form

$$\log_e Y_\mathbf{x} = \log_e Y_\mathbf{0} + \boldsymbol{\beta}^T \mathbf{x}$$

This means that as $\mathbf{x}$ changes, the distribution of $\log_e Y_\mathbf{x}$ is just that of $\log_e Y_\mathbf{0}$ subject to a location shift of $\boldsymbol{\beta}^T \mathbf{x}$.

Thus, for example, if $\mathbf{x}_j = x_j$ denotes group membership of $k+1$ groups, where $j = 0, 1, 2, 3, \ldots, k$, then, under the accelerated lifetime model, Y_j has survival function

$$\begin{aligned} S_j(y) &= S_0[y g_2(x_j)] \\ &= S_0[\exp(\log y + \log g_2(x_j))] \\ &= S_0[\exp(\log y - \beta x_j)] \\ &= S_0^*(\log y - \beta x_j), \end{aligned}$$

where $S_0^*(y) = S_0(e^y)$ is the survival function for log failure. In summary,

$$S_j(y) = S_0^*(\log y - \beta x_j), \quad j = 0, 1, 2, \ldots, k. \tag{8.10}$$

Specifically, the survival functions for 3 groups with corresponding covariate values $0, 1$ and 2 are

$$\begin{aligned} S_0(y) &= S_0^*(\log y) \\ S_1(y) &= S_0^*(\log y - \beta) \\ S_2(y) &= S_0^*(\log y - 2\beta). \end{aligned}$$

This result shows that if the groups are different, then $\beta \neq 0$ and graphs of the survival functions for each group against $\log y$ will be horizontally shifted copies of the baseline survival function S_0^* on the log scale. This may be detected in a diagnostic plot if the sample size n_j in group j is large enough.

Diagnostic 1: For accelerated lifetime models.

Recall from Definition 4.2 that an empirical survivor plot for n data points exhibits the points plotted midway in the gaps created by the jumps in the empirical survivor function: the plot consists of the points

$$(y_{(i)}, 1 - p_i), \quad i = 1, 2, 3, \ldots, n,$$

for plotting positions

$$p_i = \frac{i - \frac{1}{2}}{n}.$$

This is designed to give an impression of the shape of the survivor function. We adapt such plots now to displaying the different group survival functions.

As before, suppose that there are $k+1$ groups and n_j data points in group j for $j = 0, 1, 2, \ldots, k$. Let p_{ij} be the plotting positions for group j; $y_{(i)j}$ the ith ranked data point in group j. By (8.10) the plot of points

$$(\log_e y_{(i)j}, 1 - p_{ij}), \quad i = 1, 2, \ldots, n_j, \quad j = 0, 1, 2, \ldots, k,$$

will show location shifted versions of the baseline survival function when an accelerated lifetimes model is appropriate. This plot is also appropriate for right-censored data when the plotting positions are chosen in terms of the product-limit estimator according to (6.4).

Diagnostic 2: For accelerated lifetime models.
Compare the standard deviations of the data in each group (or, when enough data permits, the standard deviations at each covariate level). This allows us to determine the validity of the constancy of σ in the log-linear model

$$\log_e Y_\mathbf{x} = \mu(\mathbf{x}) + \sigma Z.$$

We now examine a diagnostic check for the proportional hazards model. On taking logarithms in the proportional hazards model

$$S_\mathbf{x}(y) = [S_0(y)]^{g_1(\mathbf{x})},$$

it follows immediately that

$$\log[-\log S_\mathbf{x}(y)] = \log g_1(\mathbf{x}) + \log[-\log S_0(y)].$$

As before, if $\mathbf{x}_j = x_j$ denotes group membership of $k+1$ groups, where $j = 0, 1, 2, 3, \ldots, k$, then, under the proportional hazards model, the survival times Y_i and Y_j for groups i and j (with survival functions $S_i(y)$ and $S_j(y)$) satisfy $\log[-\log S_i(y)] = \log g_1(\mathbf{x}_i) + \log[-\log S_0(y)]$ and $\log[-\log S_j(y)] = \log g_1(\mathbf{x}_j) + \log[-\log S_0(y)]$. The effect of the baseline survival function is removed by subtracting these two equations:

$$\log[-\log S_i(y)] - \log[-\log S_j(y)] = \log \frac{g_1(\mathbf{x}_i)}{g_1(\mathbf{x}_j)}. \quad (8.11)$$

Therefore, the graphs $t = \log[-\log S_i(y)]$ and $t = \log[-\log S_j(y)]$ against y should differ vertically at each value of y by the constant

$$\log \frac{g_1(\mathbf{x}_i)}{g_1(\mathbf{x}_j)}.$$

Diagnostic 3: For proportional hazards models.
With plotting positions defined in Diagnostic 1, by (8.11), the plot of points

$$(y_{(i)j}, \log[-\log(1 - p_{ij})]), \quad i = 1, 2, \ldots, n_j, \quad j = 0, 1, 2, \ldots, k,$$

will show vertically shifted curves when a proportional hazards model is correct. This plot is appropriate for right-censored data when the plotting positions are chosen in terms of the product-limit estimator according to (6.4). These curves are straight lines when $\log y_{(i)j}$ (rather than $y_{(i)j}$) is plotted on the horizontal axis and the Weibull model fits the data well. Such vertically shifted straight lines correspond equally to horizontally shifted straight lines when S_0^* is straightened out in Diagnostic 1 by plotting $\log[-\log(1-p_{ij})]$ in place of $1-p_{ij}$. This coincidence of parallel straight lines (when viewed from the perspectives of both horizontal separation, Diagnostic 1 — and vertical separation, Diagnostic 3) demonstrates that the effect of accelerated lifetime models and proportional hazards models is the same in the case of Weibull data.

■ Example 8.3
Failure times for steel specimens at four stress levels

McCool (1980) gives data for four independent rolling contact fatigue tests on hardened steel specimens at four different stress levels (in $\text{psi}^2/10^6$). We apply the diagnostic checks to the data using MINITAB for simplicity.

\multicolumn{4}{c}{Failure at stress levels}
S1	S2	S3	S4
0.87	0.99	1.09	1.18
1.67	0.012	0.80	0.073
2.20	0.180	1.00	0.098
2.51	0.200	1.37	0.117
3.00	0.240	2.25	0.135
2.90	0.260	2.95	0.175
4.70	0.320	3.70	0.262
7.53	0.320	6.07	0.270
14.70	0.420	6.65	0.350
27.80	0.440	7.05	0.386
37.40	0.880	7.37	0.456

(*Source*: McCool, J.I. (1980). Confidence limits for Weibull regression with censored data. *IEEE Transaction in Reliability*, R29, 145–50.)

The different stress levels are labelled S1–S4. In this example we examine the possibility that the failure times at different stress levels may be modelled according to Weibull distributions. In this case is the Weibull scale parameter the same for all four stress levels?

The Diagnostic 1 plot in Figure 8.1 plots $\log_e$ of the failure times at the respective stress levels S1–S4. The vertical axis shows the plotting positions $1-p_{ij}$ for each level. Since there are ten data points in each level, the plotting positions do not depend on j, the group index:

$$1 - p_{ij} = 1 - p_i = 1 - \frac{i - \frac{1}{2}}{10}, \quad i = 1, 2, 3, \ldots, 10,$$

may be written as: 0.95, 0.85, 0.75, 0.65, 0.55, 0.45, 0.35, 0.25, 0.15, 0.05.

DIAGNOSTICS FOR CHOOSING BETWEEN MODELS 153

Figure 8.1 dramatically shows the effect of horizontally shifted distributions, although there is some curvature occurring in the extremes of each group. Notationally, S1–S4 are represented respectively by the square, circle, triangle and diamond.

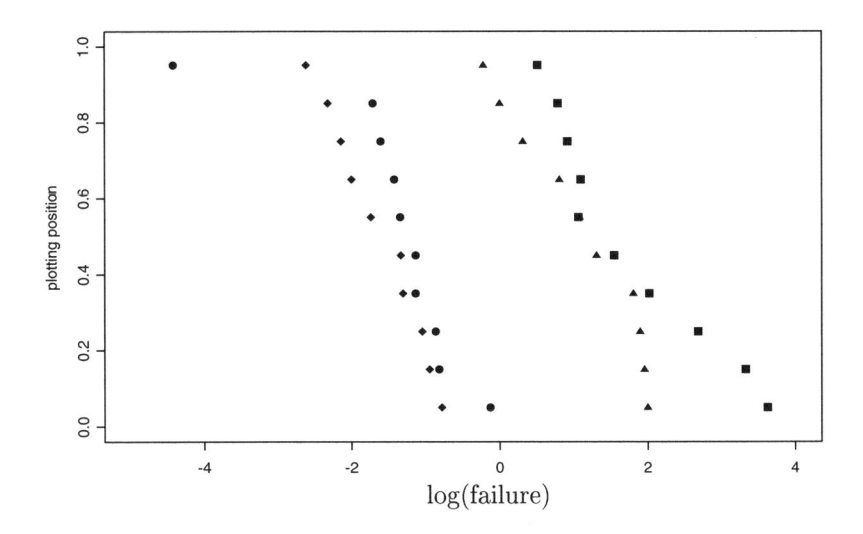

Figure 8.1 *Diagnostic 1 plot of plotting positions* $1 - p_{ij}$ *against log(failure) for the different stress levels denoted by different symbols.*

For a Diagnostic 2 analysis: on the log(failure) scale, the standard deviations of the four groups are 1.11, 1.14 (with possible outlier), 0.84 and 0.63. These show some fluctuation, although the group sample sizes are small. Remember that the log-linear model assumes constant standard deviation. Therefore, a variance stabilising transformation may be necessary before proceeding with fitting an accelerated lifetime model.

Finally, Diagnostic 3 is plotted in Figure 8.2. This plot would show vertically shifted parallel straight lines if a proportional hazards model utilising the Weibull distribution is appropriate. There is definitely a sense of straight lines; but because they are standing on different supports on the horizontal axis, it is difficult to tell whether there is constant vertical separation between two groups at a fixed value of $\log_e y$. At a simple interpretation, the four side-by-side plots are like straightened-out plots of the empirical survivor function. If we accept a sense of linearity in these plots, possibly they have different slopes. This would mean that the scale parameter of any fitted Weibull model may differ from group to group and hence be covariate dependent.

Overall, a Weibull model is reasonable, though not perfect. As Theorem 8.1

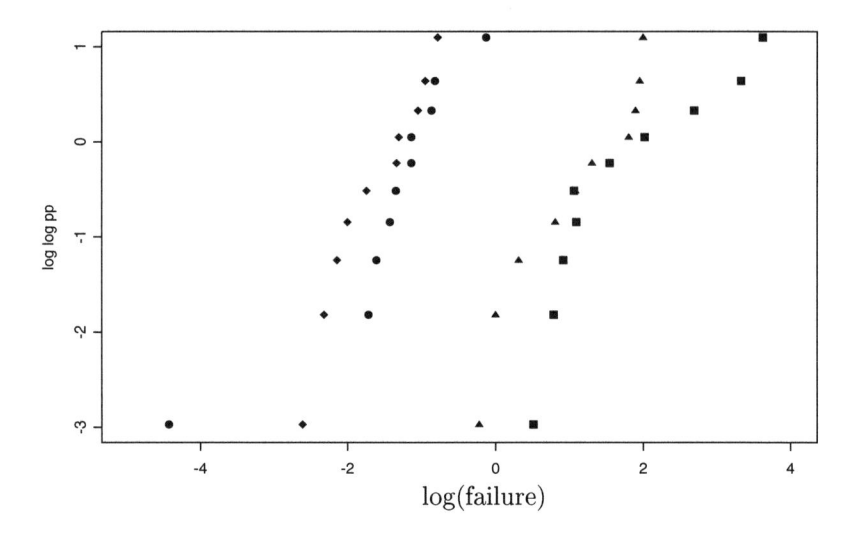

Figure 8.2 *Diagnostic 3 plot of log log plotting positions* $\log_e[-\log_e(1-p_{ij})]$ *against* log(failure) *for the different stress levels denoted by different symbols.*

suggests, such a model may be accommodated by both the proportional hazard and accelerated lifetime approaches.

□

8.6 Model fitting

We have discussed diagnostic techniques to assist in choosing which model is best for our data.

We now briefly indicate that the method of maximum likelihood may be used to find parameter estimates in our chosen models. The likelihood equations usually involve numerical solution by methods such as Newton-Rhapson. Such methods have been discussed in the previous chapter. For more specialised details we refer the reader to Lawless (1982), Kalbfleisch and Prentice (1980) and Crowder et al. (1991).

For example, to fit an accelerated lifetime model to Weibull data, we utilise

$$Z = \frac{\log_e Y_\mathbf{x} - u(\mathbf{x})}{b} \sim extremevalue(0,1)$$

from (8.8). In this case, the results of the previous chapter indicate that the log-likelihood is

$$l(u,b) = \log_e L(u,b) = -r\log_e b + \sum_{i=1}^n \delta_i(\frac{z_i - u}{b}) - \sum_{i=1}^n \exp(\frac{z_i - u}{b}),$$

MODEL FITTING

where $r = \sum_{i=1}^{n} \delta_i$ and $u = u(\mathbf{x}) = \beta^T \mathbf{x}$. Numerical MLE methods are then used to estimate b and the vector of parameters β, using standard statistical software packages that have these procedures pre-programmed. For example, the analysis in Example 8.4 is in MINITAB, but could equally well have been performed using a whole range of software (e.g. S-PLUS, SAS or STATIS-TICA).

What becomes important in fitting such models in software is an understanding of the structural implications of the model and how this structure is entered into the software. Diagnostic plots and procedures certainly assist with model selection.

■ Example 8.4
On fitting an accelerated failure time model for stressed steel specimens: Example 8.3 continued

McCool (1980) gives data for four independent rolling contact fatigue tests on hardened steel specimens at four different stress levels (in $\text{psi}^2/10^6$). In the investigative diagnostic plots of the previous section, we looked at the inclusion of the different stress levels as representing different levels of a factor or group, and accordingly examined the role of group membership as a covariate. We could look further to incorporating the numerical value of the stress level as a covariate in the modelling process. Lawless (1982) suggests from engineering considerations that if v_i denotes the ith stress level, then the modelled Weibull scale parameter may incorporate this covariate through the power law

$$w(v_i) = cv_i^r, \quad i = 1, 2, 3, 4.$$

This would entail fitting the accelerated lifetime model

$$Y_v \sim Weibull(\alpha, w(v))$$

with survival function

$$S_v(y) = \exp[-(\frac{y}{w(v)})^\alpha].$$

It is easy to place this power law in the exponential format discussed in Section 8.4, for

$$w(v_i) = cv_i^r = \exp(\beta_0 + \beta_1 \log_e v_i), \quad i = 1, 2, 3, 4, \quad (\beta_0 = \log_e c, \ \beta_1 = r).$$

This turns our accelerated lifetime model into the standard

$$Y_x \sim Weibull(\alpha, e^{\beta^T \mathbf{x}});$$

but based on $x = \log_e v$, the logarithm of the stress levels. In linear form, we therefore fit the model

$$\log_e Y_\mathbf{x} = \mu_0 + \beta^T \mathbf{x} + \sigma Z,$$

where $Z \sim extremevalue(0, 1)$, to the data using MINITAB. In this case, the covariate $\mathbf{X} = \log_e$ (stress level) with values $\log_e 0.87$, $\log_e 0.99$, $\log_e 1.09$ and $\log_e 1.18$ corresponding to S1–S4, respectively.

Regression with Life Data: log(failure) versus log v
Step Log-Likelihood
```
    0         -79.625
    1         -76.323
    2         -74.904
    3         -74.807
    4         -74.807
    5         -74.807
```

Response Variable: log(failure)
Censoring Information Count
Uncensored value 40
Estimation Method: Maximum Likelihood Dist: Extreme value

Regression Table

Predictor	Coef	Standard Error	Z	P	95.0% Normal CI Lower	Upper
Intercept	0.8916	0.2399	3.72	0.000	0.4215	1.3617
log v	-7.916	2.010	-3.94	0.000	-11.855	-3.977
Scale	1.3985	0.1670			1.1067	1.7672

Log-Likelihood = -74.807

The MINITAB software output traces the iterations for determining the maximum of the likelihood function l. The maximum likelihood estimates for μ_0, β and σ, and their standard errors, may be read from the output:

$$\widehat{\mu_0} = 0.8916; \quad \widehat{\beta} = -7.916; \quad \widehat{\sigma} = 1.3985.$$

The p-values point to the importance of each term in the model when the other term is present. Software output from S-PLUS shows similar results:

Call: survreg(formula = Surv(y) ~ x, dist = "extreme")
Coefficients:
```
              Value   Std. Error   z value       p
(Intercept)   0.892   0.24         3.72    0.000201
          x  -7.916   2.01        -3.94    0.000082
```

Extreme value distribution: Dispersion (scale) = 1.398459 Deg of Freedom: 40 Total; 37 Residual -2*Log-Likelihood: 150

Number of Newton-Raphson Iterations: 4
 Correlation of Coefficients:
  (Intercept)
x -0.213

□

8.7 Residual analysis

Example 8.4 shows the mechanics of a fit of an accelerated lifetime Weibull model to engineering failure time data at different stress levels. It is important to examine the residuals arising from the fit. The Cox-Snell residuals (Cox and Snell, 1968) are one very useful brand of residuals based on the cumulative hazard function, and therefore survival function.

Definition 8.3 *Let $H_\mathbf{x}$ denote the cumulative hazard of $Y_\mathbf{x}$. Let $\widehat{H}_\mathbf{x}$ denote the maximum likelihood estimate of $H_\mathbf{x}$ based on a fitted parametric regression model to $Y_\mathbf{x}$. When no censoring is present, the **Cox-Snell residuals**, $\widehat{r}_i$, are defined by*

$$\widehat{r}_i = \widehat{H}_{\mathbf{x}_i}(Y_{\mathbf{x}_i}).$$

*For cases where censoring is present, the **modified Cox-Snell residuals**, $\widehat{r}_i{'}$, are given by*

$$\widehat{r}_i{'} = 1 - \delta_i + \widehat{r}_i,$$

where δ_i is the censor time associated with $Y_{\mathbf{x}_i}$. □

Why and how would such residuals be effective? For uncensored data, if we evaluate a cumulative hazard function $H_\mathbf{x}$ at its own lifetime variable $Y_\mathbf{x}$, we obtain an observation from an exponential distribution with mean 1:

$$H_\mathbf{x}(Y_\mathbf{x}) \sim exp(1).$$

Therefore, as a first approximation, we expect the $\widehat{r}_i$ of Definition 8.3 to be exponentially distributed with mean 1 when the parametric model fits well. Since it is easy to show, by the results of Chapter 1, that for R_i i.i.d. $exp(1)$, when R_i is right-censored by t_i, it follows that $E(R_i|R_i > t_i) = t_i + 1$, $i = 1, 2, 3, \ldots, n$. This means that the modified Cox-Snell residuals have the censored points t_i among them replaced by the estimated conditional exponential mean $t_i + 1$. Again, for censored data, the modified Cox-Snell residuals should be treated as a random sample from an exponential distribution with mean 1 for model diagnostic purposes. Therefore, an **exponential probability plot** may be used as a check on the viability of the fitted model based on the observed Cox-Snell residuals. For example, Lawless (1982) suggests that if $\widehat{S}(\widehat{r}_i{'})$ is the Kaplan-Meier product-limit estimator based on $\widehat{r}_i{'}$, the plot

$$(\widehat{r}_i{'}, -\log_e \widehat{S}(\widehat{r}_i{'})), \quad i = 1, 2, 3, \ldots, n,$$

should follow a straight line if the proposed parametric regression model is effective.

Other residual plots may be constructed for the specific nature of the fitted parametric model. For example, if we fit the model

$$\log_e Y_\mathbf{x} = \mu_0 + \boldsymbol{\beta}^T \mathbf{x} + \sigma Z,$$

where $Z \sim extremevalue(0, 1)$, we may wish to check whether the observed

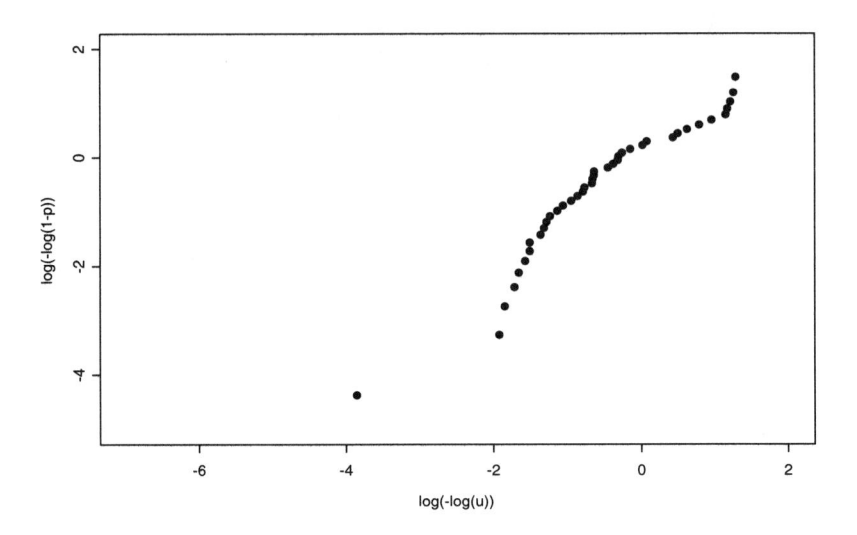

Figure 8.3 *Residual plot of log log plotting positions* $\log_e[-\log_e(1 - p_i)]$ *against* $\log_e[-\log_e(u_i)]$ *for an accelerated time model fitted to failures at different stress levels in Example 8.4.*

standardised residuals

$$l_i = \frac{\log_e Y_{\mathbf{x_i}} - \widehat{\mu_0} - \widehat{\boldsymbol{\beta}}^T \mathbf{x_i}}{\widehat{\sigma}}$$

actually follow an extreme value distribution with survival function

$$S_{extreme}(y) = \exp[-\exp(y)].$$

This may be done by transforming the l_i to scores u_i on $[0, 1]$ by

$$U_i = \widehat{S}_{\mathbf{x_i}}(Y_{\mathbf{x_i}}) \quad \text{for} \quad \widehat{S}_{\mathbf{x_i}}(y) = \exp[-(\frac{y}{e^{\widehat{\boldsymbol{\beta}}^T \mathbf{x_i}}})^{\widehat{\alpha}}].$$

It then follows that $l_i = \log[-\log u_i]$. Since, if the extreme value model fits well for standard plotting positions p_i,

$$u_{(i)} = S_{extreme}(l_{(i)}) \approx 1 - p_i,$$

then by the techniques of Chapter 4, the QQ-plot which checks for the validity of an extreme value model may be linearised by taking the ranked values of $\log_e[-\log_e u_i]$ versus $\log_e[-\log_e(1-p_i)]$. Such a plot is shown in Figure 8.3 for the model fitted in Example 8.4, indicating a departure from both linearity and the possibility of residuals following an extreme value distribution. The actual construction of the l_i values for this model, along with the associated plotting

EXERCISES

positions, is given in the S-PLUS output which follows. These represent what is plotted on the respective axes of Figure 8.3.

```
Call: survreg(formula = Surv(y) ~ x, dist = "extreme")

Coefficients:
 (Intercept)           x
   0.8915877  -7.915973

Dispersion (scale) = 1.398459 Degrees of Freedom: 40 Total; 37
Residual -2*Log-Likelihood: 149.6131

### fit.my=\hat(\mu)+\hat(\beta)*x
>  fit.my<-0.8915877 -7.915973*x
>  (log(y)-fit.my)/1.398459
 -1.0591 -0.8620 -0.7678 -0.6403 -0.6645 -0.3192  0.0178  0.4962
  0.9518  1.1639 -3.8571 -1.9206 -1.8453 -1.7149 -1.6577 -1.5092
 -1.5092 -1.3148 -1.2815 -0.7859 -0.3093 -0.1497  0.0754  0.4301
  0.6238  0.7858  1.1398  1.2050  1.2468  1.2786 -1.5722 -1.3616
 -1.2349 -1.1326 -0.9470 -0.6584 -0.6369 -0.4514 -0.3813 -0.2622

>  1-p <- 1-(c(1:40)-0.5)/40
 0.9875 0.9625 0.9375 0.9125 0.8875 0.8625 0.8375 0.8125 0.7875
 0.7625 0.7375 0.7125 0.6875 0.6625 0.6375 0.6125 0.5875 0.5625
 0.5375 0.5125 0.4875 0.4625 0.4375 0.4125 0.3875 0.3625 0.3375
 0.3125 0.2875 0.2625 0.2375 0.2125 0.1875 0.1625 0.1375 0.1125
 0.0875 0.0625 0.0375 0.0125
```

8.8 Exercises

8.1. Sketch the following baseline hazard function:

$$h_0(y) = \begin{cases} 1 & \text{if } 0 \le y \le 1 \\ 3 & \text{if } 1 < y. \end{cases}$$

Now sketch the proportional hazard which is multiplied by a factor of $g_1(x) = 2$. Compare this, on the same graph, with the accelerated lifetime hazard which is accelerated by a factor $g_2(x) = 2$.

On a separate graph, compare the survivor functions of all three hazards.

8.2. If $Z \sim extremevalue(0, 1)$ and $Y = \mu + \sigma Z$, determine the survival function of Y.

8.3. Consider the following data from Nelson (1972) showing the time to breakdown of an insulating fluid, subjected in a lifetest to various constant elevated test voltages. These data are not censored.

Failure at test voltage levels			
30 kV	32 kV	34 kV	36 kV
7.74	0.27	0.19	0.35
17.05	0.40	0.78	0.59
20.46	0.69	0.96	0.96
21.02	0.79	1.31	0.99
22.66	2.75	2.78	1.69
43.40	3.91	3.16	1.97
47.30	9.88	4.15	2.07
139.07	13.95	4.67	2.58
144.12	15.93	4.85	2.71
175.88	27.80	6.50	2.90
215.10	53.24	7.35	3.67
	82.85	8.01	3.99
	89.29	8.27	5.35
	100.58	12.06	13.77
	194.90	31.75	25.50
		32.52	
		33.91	
		36.71	
		72.89	

(*Source*: Nelson, W. (1972). Graphical analysis of accelerated life test data with the inverse power law model. *IEEE Transactions on Reliability*, R21, 2–11.)

Use appropriate diagnostic tests to examine whether a proportional hazards model or an accelerated lifetime model seems appropriate for these data. Does a Weibull model provide an adequate fit? Discuss.

8.4. Example 8.3 gives failure time data for four independent rolling contact fatigue tests on hardened steel specimens at four different stress levels (in psi^2/10^6).

If s denotes stress level, engineering theory predicts that a Weibull power law will describe the failure times Y_s at stress level s: $Y_s \sim Weibull(\alpha, ps^q)$. This means that the scale parameter is modelled as a power of the stress level, while the shape parameter remains constant.

Show how to write this as a log-linear model in the manner of (8.8). For this model, show that for suitably chosen β_0 and β_1, $u(\mathbf{x}) = \beta_0 + \beta_1 x$ where $x = \log_e s$. Thus, a power law may be fitted in a linear way.

8.5. The following data from Chambers et al. (1983) give the time to failure, measured in millions of operations, of 40 mechanical devices. Each device had two switches, A or B, which could fail; the failure of either one of these switches was fatal to the operation of the device. Any failure time

EXERCISES 161

recorded by the failure of one of the switches, say A, represents a censored observation on the failure time from switch B. In three cases, labelled '-', the device did not fail.

Failure time	Failure mode A	Failure mode B	Failure time	Failure mode A	Failure mode B
1.151		x	2.119		x
1.170		x	2.135	x	
1.248		x	2.197	x	
1.331		x	2.199		x
1.381		x	2.227	x	
1.499	x		2.250		x
1.508		x	2.254	x	
1.534		x	2.261		x
1.577		x	2.349		x
1.584		x	2.369	x	
1.667	x		2.547	x	
1.695	x		2.548	x	
1.710	x		2.738		x
1.955			2.794	x	
1.965	x		2.883	-	-
2.012		x	2.883	-	-
2.051		x	2.910	x	
2.076		x	3.015	x	
2.109	x		3.017	x	
2.116		x	3.793	-	-

(*Source*: Chambers, J.M., Cleveland, W.S., Kleiner, B. and Tukey, P.A. (1983). *Graphical Methods for Data Analysis*. Wadsworth: California.)

Discuss the fitting of a parametric regression model to these data with failure mode as a group covariate. Construct plots that help detect whether a proportional hazards model or an accelerated lifetime model is appropriate. Examine plots of residuals from your chosen model.

8.6. If survival time Y follows a Weibull distribution, show that the proportional hazards model and the accelerated time model coincide.

That is, show that if
$$S_0(y) = e^{-(\frac{y}{\beta})^\alpha},$$
then
$$S_0(y)^{g_1(x)} = S_0[yg_2(x)]$$
for specific functions $g_1(x)$ and $g_2(x)$.

8.7. Show that if the proportional hazards model and the accelerated time model coincide, then survival times follow a Weibull distribution.

That is, show that if
$$S_0(y)^{g_1(x)} = S_0[y g_2(x)],$$
then
$$S_0(y) = e^{-(\frac{y}{b})^a}$$
for specific values of a and b.

8.8. When increasing stress (in GPa) is applied to carbon fibres in a bundle of 1000 fibres, the bundle finally breaks. Suppose that bundles of fixed lengths $L_1, L_2, \ldots, L_n$ are used in an experiment testing the breaking strength of fibre bundles.

Suppose further that $S_1(y)$ is the survival function for a fibre bundle of length 1 unit; that is, it denotes the probability that fracture occurs beyond stress y for a bundle of length 1 unit.

In the *weakest link hypothesis* discussed by Crowder et al. (1991), it is supposed that long fibres consist of independent links, whose weakest member determines the failure of the bundle. For a bundle of length L, this leads to the model
$$S_L(y) = S_1(y)^L,$$
where $S_L(y)$ is the survival function for a bundle of length L.

Assume that Y_L is the failure time of a bundle of length L and that
$$Y_1 \sim Weibull(\alpha, y_1).$$

(a) Show that $Y_L \sim Weibull(\alpha, y_L)$, where
$$y_L = y_1 L^{-\frac{1}{\alpha}}.$$

(b) Show that the hazard for Y_L and the hazard for Y_1 are proportional.

(c) Suppose that α is known and that y_1 needs to be estimated by maximum likelihood methods.

The data is in the form (z_i, δ_i, L_i), $i = 1, 2, 3, \ldots, n$, where, for a bundle of length L_i, the observed stress value z_i is possibly right-censored; $\delta_i = 0$ if z_i is right-censored and has value 1 otherwise.

Show that the log-likelihood is given by
$$l = \sum_{i=1}^{n} \delta_i \{\log_e \alpha + (\alpha - 1) \log_e z_i + \log_e L_i - \alpha \log_e y_1\} - y_1^{-\alpha} \sum_{i=1}^{n} L_i z_i^{\alpha}$$
under the weakest link hypothesis.

Hence show that the maximum likelihood estimator of y_1 is
$$\widehat{y_1} = \left\{ \frac{1}{r} \sum_{i=1}^{n} L_i z_i^{\alpha} \right\}^{\frac{1}{\alpha}},$$
where $r = \sum_{i=1}^{n} \delta_i$.

EXERCISES

8.9. Suppose that Y_1 represents the survival function for a patient with covariate $\mathbf{x} = \mathbf{x_1}$ and Y_2 represents the survival function for a patient with covariate $\mathbf{x} = \mathbf{x_2}$. Show that for all y, under a proportional hazards model, either $P(Y_1 > y) > P(Y_2 > y)$ or $P(Y_1 > y) < P(Y_2 > y)$.

8.10. Suppose that the survival prospects for two groups of patients are to be compared.

Group 0, indicated by $x = x_0 = 0$, corresponds to the placebo, has hazard function $h_0(z)$, and has a modelled survival function

$$S_0(z) = e^{-\frac{z}{\lambda}}.$$

Group 1, indicated by $x = x_1 = 1$, corresponds to a new treatment, has hazard function $h_1(z)$, and has a modelled survival function

$$S_1(z) = e^{-\frac{\psi z}{\lambda}}.$$

(a) Find expressions for $h_0(z)$ and $h_1(z)$ in terms of λ and ψ.

(b) Show that the two groups follow the proportional hazards model

$$h_i(z) = e^{\beta x_i} h_0(z) \quad \text{for} \quad i = 0, 1$$

when $\psi = e^\beta$.

(c) Given the observed Group 0 data

$$(z_{i0}, \delta_{i0}), i = 1, 2, \ldots n_0$$

and the observed Group 1 data

$$(z_{i1}, \delta_{i1}), i = 1, 2, \ldots n_1,$$

where the $\delta_{ij} = 0$ if z_{ij} is right-censored and has value 1 otherwise, show that the joint log-likelihood for the two-sample problem is

$$l(\psi, \lambda) = -(r_0 + r_1) \log_e \lambda + r_1 \log_e \psi - \frac{1}{\lambda}(T_0 + \psi T_1),$$

where $r_i = \sum_{j=1}^{n_i} \delta_{ji}$, $i = 0, 1$, gives the total number of uncensored observations in Group i, and $T_i = \sum_{j=1}^{n_i} z_{ji}$, $i = 0, 1$, gives the total observed lifetime of individuals in Group i.

(d) Show that the maximum likelihood estimates of λ and ψ are given by

$$\widehat{\lambda} = \frac{T_0}{r_0} \quad \text{and} \quad \widehat{\psi} = \frac{r_1 T_0}{r_0 T_1}.$$

8.11. Show that, for uncensored data, if a cumulative hazard function $H_\mathbf{x}$ is evaluated at its own lifetime variable $Y_\mathbf{x}$, we obtain an observation from an exponential distribution with mean 1:

$$H_\mathbf{x}(Y_\mathbf{x}) \sim exp(1).$$

Use this to show that if $R_i = H_{\mathbf{x}_i}(Y_{\mathbf{x}_i})$, then

$$E(R_i | R_i > t_i) = t_i + 1.$$

8.12. Determine a formula for $S^{-1}(u)$, where S is the survival function of $Y \sim extremevalue(0,1)$ and $0 < u < 1$. Suppose that we wish to determine whether the residuals $l_1, l_2, \ldots, l_n$ follow an extreme value distribution. Determine exactly what is plotted in a QQ-plot for the extreme value distribution of the theoretical quantile $\xi_{p_i} = S^{-1}(1 - p_i)$ (as defined in Definition 4.3) against the sample quantile $l_{(i)}$.

8.13. Show that if the standardised residuals, l_i, resulting from fitting an accelerated time model are defined by

$$l_i = \frac{\log_e Y_{\mathbf{x_i}} - \widehat{\mu_0} - \widehat{\boldsymbol{\beta}}^T \mathbf{x_i}}{\widehat{\sigma}}$$

and

$$u_i = \widehat{S}_{\mathbf{x_i}}(Y_{\mathbf{x_i}}),$$

then the u_i and l_i are related by

$$l_i = \log[-\log u_i].$$

8.14. Fit accelerated lifetime models to the data of Example 8.3, assuming the responses follow Weibull, log-logistic, and lognormal distributions. Compare the effectiveness of the three fits by constructing probability plots of the Cox-Snell residuals resulting from the fits.

8.15. Suppose that the Weibull survival variable Y_x with survival function S_x depends on a covariate x according to

$$Y_x \sim Weibull(\alpha, e^{\beta x}).$$

(a) Write an expression for $S_x(y) = P(Y_x > y)$.
(b) Determine an expression for the baseline hazard function.
(c) Show that the hazards associated with Y_x for different values of x are proportional.
(d) Show that

$$\frac{\log_e Y_x - \beta x}{\alpha^{-1}} \sim extremevalue[0, 1].$$

Hence show that $\log_e Y_x$ follows the linear model

$$\log_e Y_x = \mu_0 + \mu_1 Z$$

for specified values of μ_0, μ_1 and Z.

8.16. Suppose that a survival variable T with covariate x has survival function $S_x^T(t)$ and hazard function $h_x^T(t)$ satisfying

$$h_x^T(t) = h_0^T(t) g(x) \ldots (*)$$

where h_0^T is the baseline hazard and g is a positive function satisfying $g(0) = 1$.

(a) Explain why (*) is a proportional hazards model.
(b) Show that if $Y = \log_e T$, then Y has survival function S_x^Y given by
$$S_x^Y(y) = S_x^T(e^y).$$
(c) Show that if $Y = \log_e T$, then Y has hazard function h_x^Y satisfying (*); that is, show that
$$h_x^Y(y) = h_0^Y(y)g(x).$$
(d) Suppose that $T \sim Weibull(2, \frac{1}{\sqrt{g(x)}})$.

(i.) Show that T has hazard function h_x^T satisfying (*).
(ii.) Show that the survival function $S_x^T(t)$ follows an accelerated lifetime model.

CHAPTER 9

Cox Proportional Hazards

In this chapter we continue working with the proportional hazards model of (8.2) in the previous chapter:

$$h_{\mathbf{x}}(y) = h_0(y)e^{\boldsymbol{\beta}^T \mathbf{x}}, \tag{9.1}$$

where $\mathbf{x}^T = (x_1, x_2, \ldots, x_p)$, $\boldsymbol{\beta}^T = (\beta_1, \beta_2, \ldots, \beta_p)$ and $h_0(y) = h_\mathbf{0}(y)$ is the baseline hazard occurring when $\mathbf{x} = \mathbf{0}$. In Chapter 8 by assuming a particular form for h_0 (such as a Weibull distribution, for example), a fully parametric model for proportional hazards was obtained that could effectively model censored data. The parameters in the proportional hazards model so generated were estimated by the method of maximum likelihood.

9.1 The Cox model

There have always been many advocates for the distribution-free modelling of data, especially in medical contexts. For censored data, the distribution-free plots of the product-limit estimator are widely used and very helpful for the visual comparison of survival prospects for several groups of patients, say, from the same clinical trial. In this chapter we take a distribution-free approach in the proportional hazards context and assume no parametric form for the baseline hazard. The foundation work in this area was done by Cox (1972). This work has become a platform for building the methodology of the last 30 years. The Cox model is the most important distribution-free regression model used for the analysis of censored data.

Central to the method of Cox is the idea that the hazard function is the basis of the regression model. This is, of course, different from classical regression modelling of the log-linear accelerated time models of Chapter 8. Intuitively, hazards are a measure of imminent risk, and it is reasonable to model this effectively using covariates. Cox's remarkable achievement was that his methods allowed to first estimate $\boldsymbol{\beta}$ in (9.1) and then the baseline hazard in data that are possibly right-censored. A good historical discussion is given in Miller (1981) and Collett (1994). Importantly, the baseline hazard does not need to be first estimated before the regression coefficients are estimated; thus, estimation of the baseline hazard is deferred until after an estimate of $\boldsymbol{\beta}$ is obtained.

We begin by assuming that there are no ties in the n observed lifetimes $z_i = \min\{y_i, t_i\}$, for $i = 1, 2, \ldots, n$. Some of these observed lifetimes z_i will be censored, $z_i = t_i$, and some will be uncensored, $z_i = y_i$. Specifically, suppose that k death times are uncensored and the remaining $n - k$ are uncensored. Since the procedure for the estimation of β does not depend on h_0, Cox (1972) indicates that the censored times contribute no information toward the estimation of β since h_0 could conceivably be 0 in the intervals between the observed death times (upon which the estimation of β is based). Essentially, the censoring of an individual has the simple effect that there is one less individual 'at risk' at the censoring time, so that censored individuals drop from the risk sets as time passes. This is conceptually very different from the distribution-free classical linear regression approach followed in Chapter 10, where the censor times are intimately intricated in the estimation of β. Since we will keep the censor times out of the direct calculation of β, we begin by listing the uncensored times: with a slight abuse of notation, we order the observed lifetimes (no ties) as

$$y_{(1)} < y_{(2)} < y_{(3)} < \cdots < y_{(k)},$$

and utilise these directly in the construction of the 'Cox (conditional) likelihood'. We now give details of how this likelihood is constructed.

Let $\mathbf{x}_{(i)}$ denote the vector of covariates associated with $y_{(i)}$. The remaining $n - k$ censored times enter the likelihood through the sequential consideration of the k 'risk sets'.

Definition 9.1 *The **risk set** R_i at time $y_{(i)}$ is the set of subjects alive and under observation at time $y_{(i)}^-$, immediately prior to $y_{(i)}$, $i = 1, 2, 3, \ldots, k$.* □

We will also use the risk set notation $R_i = R[y_{(i)}]$. Definition 9.1 agrees with Definition 5.3, for the number at risk in a lifetable interval, being the number alive and under observation (both censored and uncensored) immediately at the start of the interval. In terms of the lifetable notation, N_i is the number of points in R_i. We can see this notation at work in a simple data set in Example 9.1.

We now use the concept of risk sets directly in constructing the Cox likelihood. From Chapter 1, we begin by recalling how the hazard function is constructed. In particular, we can write down the relationship between 'conditional probability of survival' and the 'hazard function': for an individual with covariate $\mathbf{x}$,

$$h_{\mathbf{x}}(y) \approx \frac{P(y < Y_{\mathbf{x}} < y + \triangle y | Y > y)}{\triangle y},$$

so that we may write the approximation

$$h_{\mathbf{x}}(y)\triangle y \approx P(y < Y_{\mathbf{x}} < y + \triangle y | Y > y) \approx P(\text{Die at } y | \text{Alive up to } y)$$

between hazard function and probability. In the Cox procedure we consider the observed death times, in turn, along with their associated risk sets.

THE COX MODEL

■ Example 9.1
Constructing risk sets

The following are the first 6 data points from the data set of Example 5.3 showing the astrocytomas survival time in months after diagnosis. Two of the data points are right-censored and four of the data points are observed deaths.

Individual	Lifetime z_i	Censoring δ_i
1	7	1
2	124	1
3	88	0
4	13	1
5	2	0
6	79	1

(*Source*: M. Staples, The Anti-Cancer Council of Victoria, Australia. *Private communication*, 1994.)

From this array of data we find with $n = 6$ and $k = 4$ that $y_{(1)} = 7$, $y_{(2)} = 13$, $y_{(3)} = 79$, and $y_{(4)} = 124$, with associated risk sets:
$R_1 = \{1, 2, 3, 4, 6\}$; $R_2 = \{2, 3, 4, 6\}$; $R_3 = \{2, 3, 6\}$; $R_4 = \{2\}$. □

We start the likelihood construction by considering each term in its product. Consider first the jth term. By considering $y_{(j)}$ along with its associated risk set R_j in terms of the probability of imminent death at $y_{(j)}$ conditional on R_j, we may write the approximation:

$P(\text{Member of } R_j \text{ with covariate } \mathbf{x}_{(j)} \text{ dies at } y_{(j)} | \text{A member of } R_j \text{ dies at } y_{(j)})$

$= \dfrac{P(\text{The individual from } R_j \text{ with covariate } \mathbf{x}_{(j)} \text{ dies at } y_{(j)})}{P(\text{A member of } R_j \text{ dies at } y_{(j)})}$

$\approx \dfrac{h_{\mathbf{x}_{(j)}}(y_{(j)}) \Delta y}{\sum_{l \in R_j} h_{\mathbf{x}_l}(y_{(j)}) \Delta y}$

$= \dfrac{h_{\mathbf{x}_{(j)}}(y_{(j)})}{\sum_{l \in R_j} h_{\mathbf{x}_l}(y_{(j)})}$

$= \dfrac{h_0(y_{(j)}) e^{\boldsymbol{\beta}^T \mathbf{x}_{(j)}}}{\sum_{l \in R_j} h_0(y_{(j)}) e^{\boldsymbol{\beta}^T \mathbf{x}_l}}$

$= \dfrac{e^{\boldsymbol{\beta}^T \mathbf{x}_{(j)}}}{\sum_{l \in R_j} e^{\boldsymbol{\beta}^T \mathbf{x}_l}}$

It is important to note in this argument, the exact stage

$$\frac{h_0(y_{(j)})e^{\beta^T x_{(j)}}}{\sum_{l \in R_j} h_0(y_{(j)})e^{\beta^T x_l}} \qquad (9.2)$$

when the proportional hazards model is inserted into the conditional probability approximation, that the baseline hazard cancels from the numerator and denominator of (9.2). This is why the estimation process is able to proceed without the baseline hazard being estimated first.

Taking the product of these approximations to conditional probabilities gives the **Cox conditional likelihood**

$$L_c(\beta) = \prod_{j=1}^{k} \frac{e^{\beta^T x_{(j)}}}{\sum_{l \in R_j} e^{\beta^T x_l}}. \qquad (9.3)$$

Notice that (9.3) is not a likelihood in the sense of Definition 7.1 because the observed lifetimes, both censored and uncensored, do not figure numerically in its calculation; in particular, the censored observations impact as they sequentially drop from the risk sets in the denominator of (9.3). The term **Cox partial likelihood** is often used to address this changed role for the observed lifetimes in the construction of L_c. Cox and later authors suggest treating the partial likelihood in (9.3) in the same manner as an ordinary likelihood in the sense that general maximum likelihood theory still applies, as is detailed in the next section. This provides an asymptotic distribution for the maximising value of β, and a hypothesis testing and confidence interval framework.

Treatment of ties

The likelihood in (9.3) is easily adjusted to accommodate a smallish number of ties. Breslow (1974) suggests using the approximation where the ties actually occur sequentially as if they were distinct. Essentially, if there are d_j repetitions of an observed death time $y_{(j)}$, then this creates d_j more terms in the product; the ties however will record possibly different covariate values; write them as

$$\mathbf{x}_{(j)_1}, \mathbf{x}_{(j)_2}, \mathbf{x}_{(j)_3}, \ldots, \mathbf{x}_{(j)_{d_j}}.$$

These appear in the numerator of L_c as

$$e^{\beta^T \mathbf{x}_{(j)_1}} e^{\beta^T \mathbf{x}_{(j)_2}} \ldots e^{\beta^T \mathbf{x}_{(j)_{d_j}}} = e^{\beta^T \sum_{l=1}^{d_j} \mathbf{x}_{(j)_l}} = e^{\beta^T \mathbf{s}_j},$$

where

$$\mathbf{s}_j = \sum_{l=1}^{d_j} \mathbf{x}_{(j)_l}$$

THE COX MODEL

is the sum of the covariate vectors for tied observed death times. Thus, the additional products appearing in (9.3) simplify to

$$L_c(\boldsymbol{\beta}) = \prod_{j=1}^{k} \frac{e^{\boldsymbol{\beta}^T \mathbf{s}_j}}{[\sum_{l \in R_j} e^{\boldsymbol{\beta}^T \mathbf{x}_l}]^{d_j}}. \qquad (9.4)$$

■ Example 9.2
Building terms in the partial likelihood

We present six data points from the Stanford Heart Transplant data of Example 5.2 to demonstrate in simple terms how each term in the Cox partial likelihood is constructed so as to incorporate an 'Age' covariate. Consider a reduced data set of $n = 6$ patients with observed lifetimes Z, censoring δ and covariate $x =$ Age in months after transplant.

Individual label	Lifetime Z	Censor δ	Age x
1	51	1	50
2	51	1	47
3	322	1	48
4	838	0	42
5	339	0	54
6	551	1	50

There are $k = 3$ individual uncensored lifetimes including one tie. Therefore 3 terms need to be constructed and their product will form the likelihood L_c.

We begin by ranking the data and constructing the risk sets, being careful to keep the original labels to define the individuals in the risk sets:

Individual label	Lifetime $Y_{(j)}$	Censor δ	Ties d	Age x	Risk set R_j
1,2	$y_{(1)} = 51$	1,1	2	50,47	$\{1,2,3,4,5,6\}$
3	$y_{(2)} = 322$	1	1	48	$\{3,4,5,6\}$
5	Censored $= 339$	0	1	54	
6	$y_{(3)} = 551$	1	1	50	$\{4,6\}$
4	Censored $= 838$	0	1	42	

The first term in the partial likelihood ($k = 1$) involves a tie. Therefore, $\mathbf{s}_1 = \sum_{l=1}^{d_1} \mathbf{x}_{(1)l} = 50 + 47$. The first term may then be written as

$$\frac{e^{(50+47)\beta}}{(e^{50\beta} + e^{47\beta} + e^{48\beta} + e^{54\beta} + e^{50\beta} + e^{42\beta})^2}.$$

For the second term ($k = 2$), the first two tied observations drop from the risk set. The second term may then be written as

$$\frac{e^{48\beta}}{e^{48\beta} + e^{54\beta} + e^{50\beta} + e^{42\beta}}.$$

For the final term in the product ($k = 3$), a censored observation drops from the risk set. The third term may then be written as

$$\frac{e^{50\beta}}{e^{50\beta} + e^{42\beta}}.$$

Therefore, the partial likelihood, $L_c(\beta)$, may be written as the product

$$[\frac{e^{(50+47)\beta}}{(e^{50\beta} + e^{47\beta} + e^{48\beta} + e^{54\beta} + e^{50\beta} + e^{42\beta})^2}][\frac{e^{48\beta}}{e^{48\beta} + e^{54\beta} + e^{50\beta} + e^{42\beta}}][\frac{e^{50\beta}}{e^{50\beta} + e^{42\beta}}].$$

This expression, even for such a small data set, requires numerical Newton-Rhapson techniques to determine the maximum. The likelihood for the entire Stanford Heart Transplant data set is considerably larger. □

9.2 Maximum likelihood procedures

Following the ideas presented in Section 7.1, maximum likelihood procedures for the estimation of β in the Cox partial likelihood, L_c, require a solution to $L'_c(\beta) = \frac{\partial}{\partial \beta} L_c(\beta) = 0$. This is usually accomplished numerically via a numerical Newton-Rhapson procedure as for the usual maximum likelihood theory. Usually, the **log-likelihood** $l_c(\beta) = \log_e L_c(\beta)$ is easier to maximise, so we define the **Cox regression estimate** $\widehat{\beta}_c$ of β by

$$\frac{\partial}{\partial \beta} l_c(\beta)|_{\beta=\widehat{\beta}_c} = \mathbf{0}.$$

The derivative of the log conditional likelihood is the **score function**:

$$u_c(\beta) = \frac{\partial}{\partial \beta} l_c(\beta) = \left(\frac{\partial l_c}{\partial \beta_1}, \frac{\partial l_c}{\partial \beta_2}, \dots, \frac{\partial l_c}{\partial \beta_p}\right)^T$$

and $u_c(\beta) = \mathbf{0}$ represents the system of equations

$$\frac{\partial}{\partial \beta_r} l_c(\beta) = 0, \quad r = 1, 2, 3 \dots, p.$$

The score function is solved and asymptotic variance estimates obtained by using the Newton-Rhapson method of scoring as described in Section 7.1. This method is particularly based on the **observed information matrix** because of the presence of censoring and the difficulty in obtaining expressions for the expected value of the derivative of the score function. Specifically, in terms of the observed information matrix, in the notation of Miller (1981),

$$i(\boldsymbol{\beta}) = -[\frac{\partial^2 l_c}{\partial \boldsymbol{\beta}^2}] = - \begin{bmatrix} \frac{\partial^2}{\partial \beta_1 \partial \beta_1} \log_e L_c(\boldsymbol{\beta}) & \cdots & \frac{\partial^2}{\partial \beta_1 \partial \beta_p} \log_e L_c(\boldsymbol{\beta}) \\ \vdots & & \vdots \\ \frac{\partial^2}{\partial \beta_p \partial \beta_1} \log_e L_c(\boldsymbol{\beta}) & \cdots & \frac{\partial^2}{\partial \beta_p \partial \beta_p} \log_e L_c(\boldsymbol{\beta}) \end{bmatrix}$$

the required solution $\widehat{\boldsymbol{\beta}}_c$ results when the iterative scheme

$$\widehat{\boldsymbol{\beta}}^1 = \widehat{\boldsymbol{\beta}}^0 + i^{-1}(\widehat{\boldsymbol{\beta}}^0) \frac{\partial}{\partial \boldsymbol{\beta}} l_c(\widehat{\boldsymbol{\beta}}^0)$$

converges after passing through a sequence of estimates $\widehat{\boldsymbol{\beta}}^0, \widehat{\boldsymbol{\beta}}^1, \widehat{\boldsymbol{\beta}}^2 \ldots$ Under regularity conditions,

$$\widehat{\boldsymbol{\beta}}_c \sim_a N(\boldsymbol{\beta}, i^{-1}(\widehat{\boldsymbol{\beta}}_c)),$$

so that the variances of parameter estimates are found on the diagonal of the inverted observed information matrix $i^{-1}(\widehat{\boldsymbol{\beta}}_c)$.

Hypothesis tests are based on this asymptotic information matrix. For testing $H_0 : \boldsymbol{\beta} = \mathbf{0}$ against suitable alternatives, the score test statistic is used:

$$u_c(\mathbf{0})^T i^{-1}(\mathbf{0}) u_c(\mathbf{0}) \sim \chi_p^2 \quad \text{under } H_0.$$

This test avoids specific calculation of $\widehat{\boldsymbol{\beta}}_c$.

9.3 p = 1 and the log-rank test

Typically, the case of $p = 1$ explanatory variables in proportional hazards modelling allows us to compare the survival prospects of two groups — perhaps a treatment group and a control group. The covariate x is used to identify which of the two groups an observation belongs to: we set

$$x = \begin{cases} 0 & \text{if observation is in Group 1} \\ 1 & \text{if observation is in Group 2.} \end{cases}$$

Then model (9.1) reduces to $h_x(y) = h_0(y)e^{\beta x}$, which is expressed in terms of the single parameter β. Let us investigate how the model equations simplify in this special case. Begin by ordering the observed death times: $y_{(1)} < y_{(2)} < y_{(3)} < \ldots < y_{(k)}$ and use the obvious notation:

n_{1j} (respectively n_{2j}) denote the numbers in the risk set R_j from Group 1 (respectively Group 2):

$$n_{1j} + n_{2j} = n_j.$$

d_{1j} (respectively d_{2j}) denote the numbers of deaths at $y_{(j)}$ from Group 1 (respectively Group 2):

$$d_{1j} + d_{2j} = d_j.$$

With this notation for the numbers at risk and the numbers observed to die, we can simplify the log-likelihood from (9.4) to obtain

$$L_c(\boldsymbol{\beta}) = \prod_{j=1}^{k} \frac{e^{\boldsymbol{\beta}^T \mathbf{s}_j}}{[\sum_{l \in R_j} e^{\boldsymbol{\beta}^T \mathbf{x}_l}]^{d_j}} = \prod_{j=1}^{k} \frac{e^{\beta d_{2j}}}{[\sum_{l \in R_j} e^{\boldsymbol{\beta}^T \mathbf{x}_l}]^{d_j}},$$

and on taking logarithms, this reduces to

$$\begin{aligned} l_c(\beta) &= \log_e L_c(\beta) \\ &= \sum_{j=1}^{k} \beta d_{2j} - \sum_{j=1}^{k} d_j \log_e [\sum_{l \in R_j} e^{\beta x_l}]. \end{aligned}$$

Therefore, the log likelihood simplifies to

$$l_c(\beta) = \beta \sum_{j=1}^{k} d_{2j} - \sum_{j=1}^{k} d_j \log_e [n_{1j} + n_{2j} e^{\beta}]. \tag{9.5}$$

■ Example 9.3
Comparing treatments for leukemia remission
Lawless (1982) gives data for comparing the remission times (in weeks) for patients subjected to two different treatments. Twenty patients were assigned to each of the two treatments; '+' denotes right-censoring. A single covariate is used to compare the two treatments:

Treatment A 1 3 3 6 7 7 10 12 14 15 18 19 22
($x = 0$)   26 28+ 29 34 40 48+ 49+

Treatment B 1 1 2 2 3 4 5 8 8 9 11 12 14 16
($x = 1$)   18 21 27+ 31 38+ 44

But what is the shape of $L_c(\beta)$? We follow the descriptive pattern of Kalbfleisch and Prentice (1980) to determine the nature of this likelihood. The Cox likelihood is the product of terms in the last column of the table below: the maximising value of β is $\hat{\beta} = 0.388$ given by the S-PLUS output which follows.

We use S-PLUS to test the hypothesis that the survival curves are the same ($\beta = 0$) for the two treatments. If z contains a vector of all 40 observed death times and δ the censor indicators, then the Cox proportional hazards model is fitted simply by menus or the command:

```
coxreg.fit<-coxreg(z,delta,x)
```

This gives the contents of the file coxreg.fit which follows after the data listing.

P = 1 AND THE LOG-RANK TEST

i	$y_{(i)}$	d_i	x uncensored	x censored	Contribution to $L_c(\beta)$
1	1	3	0 1 1		$e^{2\beta}/(20+20e^\beta)^3$
2	2	2	1 1		$e^{2\beta}/(19+18e^\beta)^2$
3	3	3	0 0 1		$e^\beta/(19+16e^\beta)^3$
4	4	1	1		$e^\beta/(17+15e^\beta)$
5	5	1	1		$e^\beta/(17+14e^\beta)$
6	6	1	0		$1/(17+13e^\beta)$
7	7	2	0 0		$1/(16+13e^\beta)^2$
8	8	2	1 1		$e^{2\beta}/(14+13e^\beta)^2$
9	9	1	1		$e^\beta/(14+11e^\beta)$
10	10	1	0		$1/(14+10e^\beta)$
11	11	1	1		$e^\beta/(13+10e^\beta)$
12	12	2	1 0		$e^\beta/(13+9e^\beta)^2$
13	14	2	1 0		$e^\beta/(12+8e^\beta)^2$
14	15	1	0		$1/(11+7e^\beta)$
15	16	1	1		$e^\beta/(10+7e^\beta)$
16	18	2	1 0		$e^\beta/(10+6e^\beta)^2$
17	19	1	0		$1/(9+5e^\beta)$
18	21	1	1		$e^\beta/(8+5e^\beta)$
19	22	1	0		$1/(8+4e^\beta)$
20	26	1	0	27(1) 28(0)	$1/(7+4e^\beta)$
21	29	1	0		$1/(5+3e^\beta)$
22	31	1	1		$e^\beta/(4+3e^\beta)$
23	34	1	0	38(1)	$1/(4+2e^\beta)$
24	40	1	0		$1/(3+e^\beta)$
25	44	1	1	48(0) 49(0)	$e^\beta/(2+e^\beta)$

```
Alive Dead Deleted
    5   35      0

        coef   exp(coef)  se(coef)    z       p
[1,]   0.388     1.47      0.341    1.14   0.255

       exp(coef)  exp(-coef)  lower .95  upper .95
[1,]     1.47       0.678       0.756      2.87

Likelihood ratio test= 1.29 on 1 df, p=0.255.
Efficient score test= 1.31 on 1 df, p=0.252.
```

The large p-value indicates no significant difference between the survival functions for the two treatments. The score test has value $\frac{u(0)^2}{i(0)} = 1.31$. □

Notice that the observed information is one-dimensional, so that

$$i(\beta) = -\frac{\partial^2 \log L_c(\beta)}{\partial \beta^2}.$$

When testing the null hypothesis $H_0 : \beta = 0$, the score function and information are first found by differentiating (9.5) once and then twice with respect to β. The results are then evaluated at $\beta = 0$ to give

$$u_c(0) = \sum_{j=1}^{k} \left(d_{2j} - \frac{d_j n_{2j}}{n_j} \right) \qquad (9.6)$$

and

$$i(0) = \sum_{j=1}^{k} \frac{d_j n_{1j} n_{2j}}{n_j^2}.$$

The **efficient score test** compares the value of

$$\frac{u_c(0)^2}{i(0)}$$

with with upper tail probabilities from tables of the χ_1^2 distribution.

The expression for the score function has a built-in structure which becomes more evident when we re-express (9.6) in a different way: $u_c(0)$ may be written in the form

$$u_c(0) = \sum_{j=1}^{k} \left(d_{2j} - n_{2j} \frac{d_j}{n_j} \right), \qquad (9.7)$$

which follows the pattern

$$u_c(0) = \sum_{j=1}^{k} \left(\text{observed Gp 2 deaths at } y_{(i)} - \text{expected Gp 2 deaths at } y_{(i)} \right).$$

This is because d_{2j} has a hypergeometric distribution with expected value (being the expected number of deaths in Group 2 at $y_{(j)}$)

$$n_{2j} \frac{d_j}{n_j}.$$

The test statistic

$$\frac{u_c(0)}{\sqrt{i(0)}} \sim N(0,1)$$

based on (9.7) is the classical **Mantel-Haenszel Test** or **Log-rank test** for comparing the survival prospects for two groups through $H_0 : \beta = 0$ or H_0: the hazard functions of the two groups are the same, or H_0: the survival functions of the two groups are the same. [See Mantel and Haenszel (1959).]

Further, d_{2j} has a hypergeometric distribution with variance
$$\frac{d_j n_{1j} n_{2j} (n_j - d_j)}{n_j^2 (n_j - 1)},$$
and approximate 'binomial' variance
$$\frac{d_j n_{1j} n_{2j}}{n_j^2}.$$

(The approximation is exact when there are no tied death times and $d_j = 1$.) Then $i(0)$ is the sum of such variances; because the observed death times are independent, the variance of the sum in (9.7) is the sum of variances. Therefore, the Cox proportional hazards procedures is essentially equivalent to the Log-rank test when comparing the survival prospects of two groups. Generally, $i(0)$ will tend to overestimate the log-rank test variance.

9.4 Cox model fitting

An extensive discussion of model fitting is given in Collett (1994). Included in this discussion are the standard regression techniques for using **dummy variable covariates** to define a **factor model** for incorporating a factor A with a levels. For example, in the analysis of cancer patients in a clinical trial, we may want to incorporate the factor $A =$ 'grade of tumor' into the proportional hazards model. If there are three descriptive levels of this factor, then the proportional hazards model would need to include dummy covariates X_2 and X_3. If α_j represents the effect of Level j of the factor for $j = 1, 2, 3$, then X_2 and X_3 could have codings:

Level 1 Factor A has $X_2 = 0$ and $X_3 = 0$;
Level 2 Factor A has $X_2 = 1$ and $X_3 = 0$;
Level 3 Factor A has $X_2 = 0$ and $X_3 = 1$.

The corresponding hazards for an individual with

Factor A at level 1 has hazard $h_1(y) = h_0(y)$;
Factor A at level 2 has hazard $h_2(y) = h_0(y)e^{\alpha_2}$;
Factor A at level 3 has hazard $h_3(y) = h_0(y)e^{\alpha_3}$.

Clearly, the adoption of $\alpha_1 = 0$, allows us to identify the baseline hazard as representing an individual for which Factor A is at Level 1. Generally, an individual with vector of covariates $\mathbf{x} = (x_1, x_2)^T$, used to identify the three levels of Factor A, has hazard

$$h_\mathbf{x}(y) = h_0(y)e^{\alpha_2 x_2 + \alpha_3 x_3}. \tag{9.8}$$

Again, following standard regression techniques for the inclusion of two factors A with a levels and B with b levels in the proportional hazards model, $(a-1)(b-1)$ covariates are required. See, for example, descriptions of model fitting in standard regression texts such as Kleinbaum et al. (1988).

■ Example 9.4
Viral positivity data

Loughin and Koehler (1997) discuss bootstrapping techniques for proportional hazards models using data originally analysed by Wei et al. (1989). We will utilise only the first margin of these multivariate data to illustrate the inclusion of a factor with three levels in the proportional hazards structure. The data arose from a clinical trial designed to test the effectiveness of two dosage levels of the drug ribavirin as a retroviral treatment of AIDS patients. The data show the results for 36 patients randomly assigned to one of three treatment groups: placebo, low or high dose of the drug. Random right-censoring in these data was caused in part by sample contamination.

x_1	x_2	Y	δ	x_1	x_2	Y	δ
0	0	9	1	1	0	28	0
1	0	6	1	0	0	4	1
0	1	21	1	1	0	15	1
0	1	13	1	0	1	11	1
1	0	31	1	0	1	27	0
0	1	16	1	0	1	14	1
1	0	16	1	0	1	8	1
0	0	4	1	1	0	18	1
0	0	6	1	0	0	9	1
0	1	3	1	0	1	8	1
0	0	10	1	0	0	9	1
1	0	27	0	0	1	8	1
1	0	7	1	0	0	6	1
0	1	21	1	0	0	9	1
0	0	15	1	1	0	8	1
0	0	3	1	0	0	9	1
1	0	28	0	0	1	19	0
0	1	7	1	1	0	4	1

(*Source*: Loughin, T. M. and Koehler, K. J. (1997). Bootstrapping regression parameters in multivariate survival analysis. *Lifetime Data Analysis*, 3, 157–77.)

Samples of blood for each patient were observed and tested each day over a month until a viral marker surpassed a certain level of tolerance, at which time the patient was declared to have a positive viral test.

Standard S-PLUS output shows the estimates $\widehat{\alpha_2} = -1.51$ for the coefficient of $x_2 =$ V1 and $\widehat{\alpha_3} = -1.03$ for the coefficient of $x_3 =$ V2 for the model (9.8). Further residual diagnostics are needed to check goodness-of-fit for the overall significant 'dose level' factor in the model.

```
        coef  exp(coef) se(coef)      z        p
V1     -1.51     0.221    0.528   -2.86   0.0043
V2     -1.03     0.358    0.455   -2.26   0.0240

       exp(coef)  exp(-coef)  lower .95  upper .95
V1        0.221       4.52       0.0785      0.623
V2        0.358       2.79       0.1466      0.874

Rsquare= 0.221 (max possible= 0.993 )
Likelihood ratio test= 8.99 on 2 df, p=0.0111
Wald test = 8.98 on 2 df,            p=0.0112
Efficient score test = 10 on 2 df,   p=0.00663
```

□

In particular, as is standard procedure for regression modelling, the interaction between two factors should only be included if first the main effects themselves are included.

The difficulties of comparing two **hierarchical models** seem more compounded in proportional hazards modelling compared to classical multivariate regression. p-values associated with individual terms in a given model are to be interpreted assuming all other terms are in the given model, so their results can be misleading and care must be exercised in interpreting the output from statistical software.

One criteria suggested by Collett (1994) for model selection is to seek the smallest (positive) value of

$$-2\log_e L(\widehat{\boldsymbol{\beta}}_c)$$

when fitting different models to the same data set. In this way, an appropriately minimal set of covariates may be fitted with good predictive capabilities.

If we write:

> **Current Model**: $\widehat{L}_1$ for the likelihood when p parameters listed as $(\beta_1, \beta_2, \ldots, \beta_p)$ are estimated in Model 1;
> **Full Model**: $\widehat{L}_2$ for the likelihood when $p + q$ parameters listed as $(\beta_1, \beta_2, \ldots, \beta_p, \beta_{p+1}, \beta_{p+2}, \ldots, \beta_{p+q})$ are estimated in Model 2, with Model 1 nested within Model 2;
> **Null Model**: $\widehat{L}_0$ for the model corresponding to all parameters being set equal to 0;

then the likelihood ratio tests for the respective current and full models are:

$$-2\log_e \frac{\widehat{L}_0}{\widehat{L}_1} \quad \text{and} \quad -2\log_e \frac{\widehat{L}_0}{\widehat{L}_2},$$

having asymptotic χ_p^2 and χ_{p+q}^2 distributions under the null hypothesis that all model parameters are set equal to 0. The difference between these likelihood

ratios
$$-2\log_e \frac{\widehat{L}_0}{\widehat{L}_2} - [-2\log_e \frac{\widehat{L}_0}{\widehat{L}_1}] = -2[\log_e \widehat{L}_1 - \log_e \widehat{L}_2],$$
has an asymptotic $\chi^2_{(p+q)-p} = \chi^2_q$ distribution which may be used for testing whether the q extra parameters in the full model are zero. The test therefore involves subtracting the likelihood ratio tests in the output from each model and comparing with the upper tail of a chi-squared distribution. Traditionally, the difference
$$D = -2[\log_e \widehat{L}_1 - \log_e \widehat{L}_2]$$
where subscript 1 denotes the current model and subscript 2 denotes the full model is termed the **deviance**. The components of D are termed **deviance residuals** and were introduced by Therneau et al. (1990).

Probability plots based on the Cox-Snell residuals
$$\widehat{r}_i = \widehat{H}_{\mathbf{x}_i}(Y_{\mathbf{x}_i}),$$
as discussed in Section 8.7, provide an appropriate visual approach to goodness-of-fit of the chosen model.

■ Example 9.5
Viral positivity data: Example 9.4 continued
When the two covariates representing factor levels labelled V1 and V2 are fitted to the viral positivity data of Example 9.4, the likelihood ratio test in the S-PLUS output has value 8.99. Fitting the individual factor levels one at a time into the model creates: for V1 alone, a likelihood ratio of 3.88; for V2 alone, a likelihood ratio of 0.22.

Therefore, the usefulness of V2 when V1 is in the model is measured by subtracting the likelihood ratios: $8.99 - 3.88 = 5.11$ (p-value < 0.05).

Similarly, the usefulness of V1 when V2 is in the model is measured by subtracting the likelihood ratios: $8.99 - 0.22 = 8.77$ (p-value < 0.05). We are able to conclude that each level of the viral dose factor makes a significant contribution to the model. □

9.5 Estimating baseline survival

Once $\boldsymbol{\beta}$ has been estimated through maximum likelihood, then the hazard function for an individual with covariate $\mathbf{x}$ is estimated via
$$\widehat{h}_{\mathbf{x}}(y) = \widehat{h}_0(y) e^{\widehat{\boldsymbol{\beta}}_c^T \mathbf{x}},$$
where $\widehat{\boldsymbol{\beta}}_c$ is the Cox regression estimator of $\boldsymbol{\beta}$. This requires that the baseline hazard $h_0(y)$ must first be estimated. Equivalently, the baseline survival function $S_0(y)$ may be estimated by $\widehat{S}_0(y)$ using maximum likelihood methods as outlined by Kalbfleisch and Prentice (1973). Then the survival function $S_{\mathbf{x}}(y)$ with covariate $\mathbf{x}$ is estimated via
$$\widehat{S}_{\mathbf{x}}(y) = \widehat{S}_0(y)^{\exp(\widehat{\boldsymbol{\beta}}_c^T \mathbf{x})}.$$

The solutions to the maximum likelihood equations may be summarised as follows: For each $j = 1, 2, 3, \ldots, k$, let D_j denote the set of all d_j individuals who die at the jth death time $y_{(j)}$. Then

$$\widehat{h_0}(y_{(j)}) = 1 - \widehat{\xi}_j,$$

where $\widehat{\xi}_j$ is the solution to the equation

$$\sum_{l \in D_j} \frac{\exp(\widehat{\boldsymbol{\beta}}_c^T \mathbf{x}_l)}{1 - \widehat{\xi}_j^{\exp(\widehat{\boldsymbol{\beta}}_c^T \mathbf{x}_l)}} = \sum_{l \in R_j} \exp(\widehat{\boldsymbol{\beta}}_c^T \mathbf{x}_l), \qquad (9.9)$$

for $j = 1, 2, \ldots, k$. Generally, this equation requires numerical integration to solve for $\widehat{\xi}_j$. However, the situation is dramatically simplified when there are no ties at the jth death time, in which case the set D_j contains one member. Then, (9.9) simplifies to

$$\frac{\exp(\widehat{\boldsymbol{\beta}}_c^T \mathbf{x}_{(j)})}{1 - \widehat{\xi}_j^{\exp(\widehat{\boldsymbol{\beta}}_c^T \mathbf{x}_{(j)})}} = \sum_{l \in R_j} \exp(\widehat{\boldsymbol{\beta}}_c^T \mathbf{x}_l), \qquad (9.10)$$

which yields the solution

$$\widehat{\xi}_j = \left(1 - \frac{\exp(\widehat{\boldsymbol{\beta}}_c^T \mathbf{x}_{(j)})}{\sum_{l \in R_j} \exp(\widehat{\boldsymbol{\beta}}_c^T \mathbf{x}_l)}\right)^{\exp(-\widehat{\boldsymbol{\beta}}_c^T \mathbf{x}_{(j)})}$$

Finally,

$$\widehat{S}_0(y) = \prod_{j=1}^{u} \widehat{\xi}_j,$$

for $y_{(u)} \leq y < y_{(u+1)}$, $u = 1, 2, \ldots, k-1$. This estimation process assumes that the hazard of death is constant between adjacent death times. The values of $\widehat{\xi}_j$ may be found explicitly if $d_j = 1$ for all j, but the likelihood equations require iterative solution if $d_j > 1$ for some j. An excellent discussion and examples may be found in Collett (1994), which provides a thorough introduction to fitting proportional hazards models in the medical context.

9.6 Exercises

9.1. Simple proportional hazards models allow us to compare the survival prospects of two groups — perhaps a treatment group and a control group. The covariate x is used to identify which of the two groups an observation belongs to: we set

$$x = \begin{cases} 0 & \text{if observation is in Group 1} \\ 1 & \text{if observation is in Group 2.} \end{cases}$$

Then $h_x(y) = h_0(y)e^{\beta x}$, which is expressed in terms of the single parameter β.

Derive expressions for the hazard functions $h_1(y)$ and $h_2(y)$ for the two groups. If the corresponding survival functions are $S_1(y)$ and $S_2(y)$, show that
$$S_2(y) = S_1(y)^{exp\beta}.$$
Hence show that, under the proportional hazards model, the test $H_0 : \beta = 0$ is equivalent to the test for equality of the survival functions of the two groups.

9.2. Consider the following severe viral hepatitis data from a clinical trial in which a steroid (Group I) and control group (Group II) were compared. Lifetimes in weeks were measured as:

Group I:	1(3)	1+	4+	5	7	8	10
	10+	12+	16+(3)				
Group II:	1+	2+	3(2)	3+	5+(2)	16+(8)	

Here, the notation 4+ indicates an observed right-censored time at 4 weeks and frequencies are given in parentheses. Group I has 14 data points and Group II has 15 data points.

Suppose that Group membership is identified by the explanatory variable x, which assumes the value 1 for Group II and 0 for Group I.

(a) Construct a chart showing the ranked observed lifetimes along with censoring and group membership information.

(b) Construct the risk sets $\mathcal{R}_i$ associated with the ith largest observed lifetime.

(c) For these data, construct the Cox partial likelihood $L_c(\beta)$ for the inclusion of group membership into the analysis of survival time through a parameter β.

(d) Perform a test for the equality of the survival functions for Group I and Group II.

9.3. The following data give the observed times, z, in days, from injection with carcinogen to cancer mortality in rats. The survival times were subject to right-censoring.

Individual	z_i	x_i	δ_i
1	282	1	1
2	204	0	0
3	228	0	1
4	186	1	0
5	190	1	1
6	206	1	1
7	143	0	1

EXERCISES

Here, in the usual notation, $z_i = \min\{y_i, t_i\}$ where t_i is the censor time associated with y_i and the censor indicator δ_i returns a value 0 if z_i is censored and is 1 otherwise.

Two groups, distinguished by a pre-treatment regime, are identified by the explanatory variable x, which assumes the value 1 for Group 2 and 0 for Group 1.

(a) Which of the seven censor times $t_1, t_2, \ldots, t_7$ are known?
    State the values of those which are known.
(b) Which of the seven survival times $y_1, y_2, \ldots, y_7$ are known?
    State the values of those which are known.
(c) For this data set, construct the Cox conditional likelihood $L_c(\beta)$ for the inclusion of group membership into the analysis of survival time through a parameter β, where, in the usual notation,

$$L_c(\beta) = \prod_{i=1}^{k} \frac{e^{x_{(i)}\beta}}{\sum_{l \in R_i} e^{x_l \beta}},$$

where R_i is the risk set associated with the ith largest survival time.

(d) Show that the parameter β may be estimated as the solution to the equation

$$2 - \frac{4e^\beta}{4e^\beta + 3} - \frac{3e^\beta}{3e^\beta + 2} - \frac{2e^\beta}{2e^\beta + 1} - \frac{e^\beta}{e^\beta + 1} = 0.$$

(e) Outline how you would test for the equality of the survival functions for Group 1 and Group 2.

9.4. Consider the following data points from the Stanford Heart Transplant data set:

Mismatch X	Survival Y
2.09	54
0.36	127+
0.60	297
1.44	389+
0.91	1536+

The mismatch score is a numerical measure of how closely 'aligned' the donor tissue and the recipient tissue are. As usual, + denotes a right-censored observation.

(a) Write down the proportional hazards model for incorporating the Mismatch Score into the estimation of survival time through a parameter β.

(b) For this data set, construct the Cox partial likelihood $L_c(\beta)$, where, in the usual notation,

$$L_c(\beta) = \prod_{i=1}^{k} \frac{e^{x_{(i)}\beta}}{\sum_{l \in \mathcal{R}_i} e^{x_l \beta}}$$

(c) Test the hypothesis $H_0 : \beta = 1$ in the proportional hazards model by assuming that the statistic

$$\frac{u_c(1)^2}{i(1)},$$

where

$$u_c(\beta) = \frac{d}{d\beta} \log_e L_c(\beta) \quad \text{and} \quad i(\beta) = -\frac{d^2}{d\beta^2} \log_e L_c(\beta),$$

has an approximate chi-squared distribution with 1 degree of freedom when H_0 is true.

9.5. Shouki and Pause (1999) report data from a study of hormonal therapy for treatment of cattle with cystic ovarian disease. Two groups of cattle were randomised to hormonal treatment and one group received placebo. In this context, the 'survival time' of interest was the time to complete disappearance of all cysts; right-censoring occurs (denoted by +) when not all the cysts disappeared by the end of the study.

Cystic ovary data		
Treatment 1	Treatment 2	Placebo
4	7	19
6	12	24
8	15	18+
8	16	20+
9	18	22+
10	22+	27+
12		30+

(Source: Shouki, M.M. and Pause, C.A. (1999). Statistical Methods for Health Sciences. CRC Press: London.)

(a) Fit a Cox proportional hazards model to these data and comment on the results. What checks can be done to check the underlying assumptions concerning proportional hazards in this example?

(b) Give estimates of the fitted survival functions for each group as powers of the baseline survival function $\widehat{S_0}(y)$.

9.6. Show that when testing the null hypothesis $H_0 : \beta = 0$, in the Cox proportional hazards model when there is one parameter β, the information and score function evaluated at 0 are given by

$$u(0) = \sum_{j=1}^{k} \left(d_{2j} - \frac{d_j n_{2j}}{n_j} \right)$$

and

$$i(0) = \sum_{j=1}^{k} \frac{d_j n_{1j} n_{2j}}{n_j^2}.$$

9.7. Show that if we set $\widehat{\beta}_c = \mathbf{0}$, then the estimator

$$\widehat{S_0}(y) = \prod_{j=1}^{u} \widehat{\xi}_j,$$

[for $y_{(u)} \leq y < y_{(u+1)}$, $u = 1, 2, \ldots, k-1$] reduces to the product-limit estimator for the baseline survival function based on a single (combined) sample.

9.8.(a) Suppose that Y is an absolutely continuous survival random variable with survival function S and cumulative hazard function

$$H(u) = -\log_e S(u).$$

(i.) Show that if we set $W = H(Y)$, then W has an exponential distribution with mean $\beta = 1$.
(ii.) For any fixed value $t > 0$, use (i.) to show that $E(W|W > t) = t + 1$.

(b) The Cox-Snell residuals for the Cox proportional hazards model are defined by

$$r_i = e^{\widehat{\beta} x_i} \widehat{H}_0(y_i),$$

where the cumulative baseline hazard H_0 is estimated from the fitted Cox Proportional Hazards model in which x_i is the covariate associated with lifetime y_i through the estimated parameter $\widehat{\beta}$.

(i.) If there is no censoring, what probability distribution should the observed Cox-Snell residuals follow? Explain using the results of (a). What type of diagnostic plot could be used to assess goodness of fit using the observed Cox-Snell residuals?
(ii.) In order to account for right-censoring in the observed data, the Modified Cox-Snell residuals are defined by

$$r'_i = 1 - \delta_i + r_i.$$

Using the results of (a), explain why these should follow the same distribution as r_i, even when right-censoring is present.

9.9. The Stanford Heart Transplant data listed in Example 5.2 gives observations on 'Days', being survival time (in days) after transplant; also listed is the right-censoring information for each patient under the heading 'Cens'; age of patient 'Age'; mismatch score, 'T5', a measure of how close alignment of the donor and recipient tissue.

(a) Fit a Cox proportional hazards model to these data so as to include 'Age' and 'T5' as covariates contributing to the model.

(b) Construct a residual plot of the Cox-Snell residuals and comment on the results.

(c) Compare the model fitted in (a) with the nested models that fit 'Age' and 'T5', respectively. Does it appear that 'Age' is the only significant covariate? Explain.

9.10. (**Breslow's estimator of the baseline hazard**). Breslow (1974) began with the usual notation for the Cox model.

(a) Show that if $\widehat{\xi}_j$ is near 1, then

$$\widehat{\xi}_j^{\exp(\widehat{\boldsymbol{\beta}}_c^T \mathbf{x}_l)} \approx 1 + (\log_e \widehat{\xi}_j) \exp(\widehat{\boldsymbol{\beta}}_c^T \mathbf{x}_l).$$

(b) Substitute the result of (a) into (9.9) to show that

$$\widehat{\xi}_j = \exp\left(\frac{-d_j}{\sum_{l \in R_j} \exp(\widehat{\boldsymbol{\beta}}_c^T \mathbf{x}_l)}\right).$$

(c) Hence provide an expression for $\widehat{S}_0(y)$ and show that

$$\widehat{H}_0(y) = -\log_e \widehat{S}_0(y) = \sum_{j=1}^{u} \left(\frac{d_j}{\sum_{l \in R_j} \exp(\widehat{\boldsymbol{\beta}}_c^T \mathbf{x}_l)}\right)$$

for $y_{(u)} \leq y < y_{(u+1)}$, $u = 1, 2, \ldots, k-1$.

CHAPTER 10

Linear Regression with Censored Data

10.1 Introduction

In simple linear regression where the response variable has been subject to right-censoring, it is unwise to place the same statistical credence to all points in the scatterplot of response against covariate. Many methods that are based on traditional ideas of linear regression (as opposed to proportional hazards) have been designed specifically to account for the effect of various censoring patterns. Some of these regression methods will be detailed in this chapter. However, it is the Buckley-James method that is principally discussed here. Interest in this method has been sustained since the first proposals took shape in Buckley and James (1979). Many of the ideas (such as renovation) described in this chapter are taken from Smith and Zhang (1995).

In the Buckley-James method for simple linear regression, to compensate for right-censoring, censored points in the scatterplot of observed data are moved vertically to create a renovated scatterplot where the bias of censoring is removed. The points are moved to estimated positions which are unbiased provided that there are large expected numbers of censored and uncensored observations across the support of the data. The Buckley-James process has certain parallels with least squares estimation which will be explored.

For simple linear regression, a scatterplot records outcomes of a response variable Y in the presence of an explanatory variable X in the form of plotted points

$$(x_i, y_i), \quad i = 1, 2, 3, \ldots, n.$$

Such plots are at the heart of regression analysis.

According to Chapter 5, the response variable is termed 'censored' when the observed response is

$$z_i = y_i \delta_i + t_i(1 - \delta_i), \quad i = 1, 2, 3, \ldots, n, \tag{10.1}$$

where δ_i is the 0–1 indicator variable indicating whether y_i (when $\delta_i = 1$ and the ith data point is uncensored) or the fixed censor time t_i (when $\delta_i = 0$ and the ith data point is censored) is observed.

We consider cases where the explanatory variable is not censored and is independent of the mechanism generating the censor times. The scatterplot of censored data consists of the plotted points

$$(x_i, z_i), \quad i = 1, 2, 3, \ldots, n,$$

with the z_i defined by (10.1). When the scatterplot is used as a guide for the effect of the explanatory variable on the response, the points are in the 'wrong place': under right-censoring, $z_i = \min\{y_i, t_i\}$, so that the censored points are lower than they should be; under left-censoring, $z_i = \max\{y_i, t_i\}$, so that the censored points are higher than they should be. In what follows in this chapter we consider right-censoring, since left-censoring is a special case corresponding to a reversal of the direction of the response axis. This means that

$$\delta_i = \begin{cases} 0 & \text{if } y_i \geq t_i \\ 1 & \text{if } y_i < t_i. \end{cases}$$

Consider the simple linear model

$$Y_i = \beta_0 + \beta_1 x_i + \epsilon_i; \ \epsilon_i \sim F; \ E(\epsilon_i) = 0; \ Var(\epsilon_i) = \sigma^2 \quad (10.2)$$

for the range of index values $i = 1, 2, 3, \ldots, n$. Here F is not specified, $\sigma^2 < \infty$ and the residual distribution has unknown survival function $S = 1 - F$. We will consider various methods for the estimation of β_0 and β_1 in this model in the next sections, concentrating particularly on the Buckley-James methodology for much of Chapter 10 and all of Chapter 11.

We begin with some classical linear regression background. Note first that in ordinary least squares, in the absence of censoring, the so-called 'least squares' estimates of β_0 and β_1 are determined as those values of b_0 and b_1 minimising the residual sum of squares

$$\sum_{i=1}^{n}(y_i - b_0 - b_1 x_i)^2. \quad (10.3)$$

Performing this minimisation using calculus, we differentiate (10.3) with respect to b_0 and b_1 and set the resulting expressions equal to 0 to obtain the **least squares normal equations**:

$$\sum_{i=1}^{n}(Y_i - b_0 - b_1 x_i) = 0$$

$$\sum_{i=1}^{n} x_i(Y_i - b_0 - b_1 x_i) = 0$$

Clearly, the estimate of β_1 may be found as the solution to the second of these equations

$$\sum_{i=1}^{n}(x_i - \bar{x})(Y_i - b_1 x_i) = 0. \quad (10.4)$$

This gives

$$\widehat{\beta_1} = \sum_{i=1}^{n} \frac{(x_i - \bar{x})Y_i}{(x_i - \bar{x})^2}.$$

Substituting this estimate back into the first of the normal equations gives

$$\widehat{\beta_0} = \bar{Y} - \widehat{\beta_1}\bar{x}. \quad (10.5)$$

Notice that this estimate of the intercept is the arithmetic mean of the residuals $y_i - \widehat{\beta_1} x_i$ and may be carried out separately after the slope has been estimated. Buckley-James estimation parallels some of this least squares process.

10.2 Miller's method

Consider the problem of estimating the parameters β_0 and β_1 in (10.2) where Y_i is right-censored by the fixed censor time t_i. One of the earliest estimation proposals was due to Miller (1976). The data are of the form (Z_i, δ_i, x_i), where $Z_i = \min\{Y_i, t_i\}$. It is assumed that the residuals based on any slope and intercept are independent of the censoring mechanism generating the t_i.

The estimation process is achieved by using the product-limit estimator of the residual distribution function. Let $v_i(b) = d\widehat{F_b}(e_i(b))$ denote the probability mass assigned to $e_i(b) = z_i - bx_i$ in the construction of the product-limit estimator $\widehat{F_b}$ based on $e_1(b), e_2(b), \ldots, e_n(b)$. Since these probability masses are determined from the residual ranking, they are invariant under translations of the **partial residuals** $e_i(b)$ and so do not depend on an intercept term.

Miller (1976) proposed locating the values of a and b which minimise the modified residual sum of squares

$$\int_{-\infty}^{\infty} u^2 d\widehat{F_b}(u) = \sum_{i=1}^{n} v_i(b)(z_i - a - bx_i)^2. \tag{10.6}$$

Notice that, if no censoring is present, $\widehat{F_b}$ is the empirical distribution function and (10.6) reverts to the usual least squares procedure outlined in the last section. The minimising procedure is done through iteration. Unfortunately, the minimum $b = \widehat{\beta_{Miller}}$ often occurs at a discontinuity, and may not be consistent for β_1. Mauro (1983) showed that even the stringent condition on the censor times — that they follow the true regression line — was not sufficient to guarantee consistency. Miller and Halpern (1982) compared performances of several censored regression estimators under simulation and found superior performance from Buckley-James estimators described in this chapter. What remains interesting is that an approach to censored regression based on modifying the residual sum of squares before minimisation did not perform as well as the Buckley-James approach where the least squares normal equations are modified before they are solved.

10.3 Koul-Susarla-Van Ryzin estimators

Consider the problem of estimating the parameters β_0 and β_1 in (10.2) where Y_i is right-censored by the random censor time T_i. That is, we observe (Z_i, δ_i, x_i), where $Z_i = \min\{Y_i, T_i\}$. The $y_1, Y_2, \ldots, Y_n$ and $T_1, T_2, \ldots, T_n$ are both sets of independent random variables, where Y_i and T_i are conditionally independent given x_i. This, in fact, allows the censoring variables to be covariate dependent. Let $G(u; x_i)$ denote the survival function of T_i.

The method of Koul, Susarla and Van Ryzin (1981) is particularly interesting because it does not require iteration to obtain an estimate of β_1. It is motivated by the identity

$$E[\frac{\delta_i Z_i}{G(Z_i, x_i)} | x_i] = \beta_0 + \beta_1 x_i,$$

which indicates that the variables

$$\frac{\delta_i Z_i}{G(Z_i, x_i)}$$

obey a linear regression similar to (10.2), except now the residuals may not be independent and identically distributed. If the survival function G for the censor times were known, this suggests a procedure which modifies the usual least squares normal equations (10.4) by replacing censored observations by 0, scaling uncensored observations y_i to

$$\frac{y_i}{G(y_i, x_i)},$$

leading to the estimate

$$\widehat{\beta_1} = \sum_{i=1}^{n} \frac{(x_i - \overline{x}) \frac{\delta_i y_i}{G(y_i, x_i)}}{(x_i - \overline{x})^2} \qquad (10.7)$$

of β_1. Following such a procedure, points are moved away from the line in order to estimate it. Assumptions such as $G(u; x_i) \equiv G(u)$ assist in the estimation of G, which is usually unknown. Miller and Halpern (1982) concluded that the Buckley-James estimator outperformed the Koul-Susarla-Van Ryzin estimator in the analysis of the Stanford Heart Transplant data.

10.4 Buckley-James regression methods

The Buckley-James method for estimating parameters in (10.2) was first proposed in the *Biometrika* paper by Buckley and James (1979). The methodology was gradually put to the test in the years that followed: Miller and Halpern (1982), Weissfeld and Schneider (1986), Heller and Simonoff (1990) and Hillis (1993) compared methodologies through simulation; James and Smith (1984) established the weak consistency of the regression estimators; Lai and Ying (1991) proposed variance estimates through a more general model; Smith and Zhang (1995) and Currie (1996) 'renovated' the scatterplot to account for censoring. In particular, papers such as Heller and Simonoff (1992) advocate the use of linear regression methods when proportional hazards models are inappropriate for regression data.

The method is simply based on an iterative solution to the usual least squares normal equations which have been modified to take account of the censoring. Certainly such a method is simple in concept, but it took many years before the properties of the parameter estimates, generated by the iterations, were well understood. The method was probably conceived by noticing that if censored points t_i in the scatterplot were relocated to their 'true' expected positions $E(Y_i|Y_i > t_i)$, the possible linearity in the scatterplot is not destroyed.

To see this, if we replace t_i in $z_i = y_i \delta_i + t_i(1 - \delta_i)$, by $E(Y_i|Y_i > t_i)$, then new response variables

$$V_i = Y_i \delta_i + E(Y_i|Y_i > t_i)(1 - \delta_i)$$

still have mean $\beta_0 + \beta_1 x_i$. This is easy to show by conditioning on the δ_i.

Theorem 10.1 *If $V_i = Y_i \delta_i + E(Y_i|Y_i > t_i)(1 - \delta_i)$, then these pseudo-random variables still follow the true regression line in the sense that*

$$E(V_i) = \beta_0 + \beta_1 x_i.$$

Proof:

$$\begin{aligned}
E(V_i) &= E(V_i|\delta_i = 1)P(\delta_i = 1) + E(V_i|\delta_i = 0)P(\delta_i = 0) \\
&= E(Y_i|\delta_i = 1)P(\delta_i = 1) + E[E(Y_i|Y_i > t_i)|\delta_i = 0]P(\delta_i = 0) \\
&= E(Y_i|\delta_i = 1)P(\delta_i = 1) + E(Y_i|Y_i > t_i)P(\delta_i = 0) \\
&= E(Y_i|\delta_i = 1)P(\delta_i = 1) + E(Y_i|\delta_i = 0)P(\delta_i = 0) \\
&= E(Y_i) \\
&= \beta_0 + \beta_1 x_i
\end{aligned}$$

$\square$

Consider the usual regression situation of estimating the parameters β_0 and β_1 in (10.2). Since, by Theorem 10.1, replacing censored points by their true conditional expected values does not bias the scatterplot, the Buckley-James method replaces censored points by their estimated conditional expected values.

Assume initially that $\beta_0 = 0$. This is equivalent to re-parametrising model (10.2) using residuals

$$\epsilon_i^1 = \epsilon_i + \beta_0$$

which have mean β_0 and variance σ^2 and satisfy

$$Y_i = \beta_1 x_i + \epsilon_i^1; \quad \epsilon_i^1 \sim F; \quad E(\epsilon_i^1) = \beta_0; \quad Var(\epsilon_i^1) = \sigma^2 \qquad (10.8)$$

This means that we may proceed with the estimation of slope β_1 first and later estimate intercept β_0 as the mean of the observed residual distribution. (For the usual least squares parallel, see equation (10.5).) Accordingly, new response variables $Y_i^*(b)$, based on a given slope b are created. They have been constructed to be of the form

$$Y_i^*(b) = bx_i + [\epsilon_i(b)\delta_i + \widehat{E_b}(\epsilon_i(b) \mid \epsilon_i(b) > c_i(b))(1 - \delta_i)], \qquad (10.9)$$

where we employ the different types of residual notation:

$$c_i(b) = t_i - bx_i;$$
$$\epsilon_i(b) = Y_i - bx_i;$$
$$E_i(b) = Z_i - bx_i = \min\{\epsilon_i(b), c_i(b)\}.$$

In (10.9), the estimated conditional expectation $\widehat{E_b}(\epsilon_i(b) \mid \epsilon_i(b) > c_i(b))$ inside the square brackets on the right-side of the equation is a linear combination of residuals: it is a weighted sum of uncensored residuals larger than $c_i(b)$. In this linear combination, the weights are determined by (10.10) through the product-limit estimator $\widehat{F} = 1 - \widehat{S}$ applied to the observed residuals $e_i(b)$, some of which are censored and some of which are uncensored. To simplify notation, we assume that these residuals have been written in rank order $(e_i(b) \equiv e_{(i)}(b))$ and that indicators, covariates and responses are re-ordered according to this ranking. Specifically,

$$\widehat{E_b}(\epsilon_i(b) \mid \epsilon_i(b) > c_i(b)) = \sum_{k=1}^{n} w_{ik}(b) e_k(b),$$

where

$$w_{ik}(b) = \begin{cases} \dfrac{d\widehat{F}(e_k(b))\delta_k(1-\delta_i)}{\widehat{S}(e_i(b))} & \text{if } k > i \\ 0 & \text{otherwise,} \end{cases} \quad (10.10)$$

involving $d\widehat{F}(e_k(b))$, the probability mass assigned by $\widehat{F}$ to $e_k(b)$. Note that, because of the ranking, $k > i$ if and only if $e_k(b) > e_i(b)$. The weights in (10.10) are based on those given by Buckley and James (1979) in their seminal paper.

Here we can ask the rhetorical question: how were the weights in (10.10) conceived of in the first place? We have insight into the answer through study of mean residual life expectancies in Chapter 1. Think in the following terms: if the true parameter were known, these weights would estimate $\frac{dF(r)}{S(l)}$ in the equation

$$E[\epsilon_i \mid \epsilon_i > l] = \int_l^\infty r \frac{dF(r)}{S(l)},$$

so that the conditional expectation $E(\epsilon_i \mid \epsilon_i > e_i(\beta))$ is *estimated* as a linear combination

$$\sum_{k=1}^{n} e_k(b) w_{ik}(b)$$

when b is near β.

Recall now (10.4), the second least squares normal equation. To create a censored version of this equation, we use the newly created $Y_i^*(b)$ instead of Y_i. Then, by definition, the solution to the modified least squares normal equation for slope

$$\sum_{i=1}^{n}(x_i - \overline{x})(Y_i^*(b) - bx_i) = 0$$

gives a Buckley-James slope estimator $\widehat{\beta_n}$ of β as an iteratively-derived value of b satisfying
$$\frac{\sum_{i=1}^{n}(x_i - \bar{x})Y_i^*(b)}{\sum_{i=1}^{n}(x_i - \bar{x})^2} - b = 0. \tag{10.11}$$
This equation has at least one 'solution' (Buckley and James, 1979) which is consistent (James and Smith, 1984). In fact, the so-called 'solution' may take the form of a **zero-crossing**. Let Φ_n denote the right-side of (10.11):
$$\Phi_n(b) = \frac{\sum_{i=1}^{n}(x_i - \bar{x})Y_i^*(b)}{\sum_{i=1}^{n}(x_i - \bar{x})^2} - b$$

Definition 10.1 *By a Buckley-James estimate of β_1 in the simple linear model (10.8) we mean a zero-crossing point $\widehat{\beta_n}$ where $\Phi_n(b)$ changes sign: that is, where*
$$\Phi_n(\widehat{\beta_n}^+)\Phi_n(\widehat{\beta_n}^-) \leq 0.$$
$\square$

Notice that if a zero-crossing is in fact an exact solution, so that $\Phi_n(\widehat{\beta_n}) \equiv 0$, then Definition 10.1 is satisfied.

Figure 10.1 shows the possible shape for a graph of the equation $y = \Phi_n(b)$ as a function of b. As was noted by Buckley and James (1979), it is immediately noticeable that the graph is piecewise linear, with jump discontinuities at the points where the residuals change rank (as b changes) in such a way so as to alter the orderings of the censored and uncensored residuals used in the calculation of the product-limit estimator. For a given sample size n, Φ_n is not necessarily monotone decreasing.

Careful reorganisation of the left-side of (10.11) establishes the linearity in b:
$$\Phi_n(b) = A_1 - bA_2,$$
where
$$A_1 = \frac{\sum_{i=1}^{n}(x_i - \bar{x})[Y_i\delta_i + \sum_{k=1}^{n}w_{ik}(b)y_k(1-\delta_i)]}{\sum_{i=1}^{n}(x_i - \bar{x})^2}$$
and
$$A_2 = \frac{\sum_{i=1}^{n}(x_i - \bar{x})[x_i\delta_i + \sum_{k=1}^{n}w_{ik}(b)x_k(1-\delta_i)]}{\sum_{i=1}^{n}(x_i - \bar{x})^2}$$
are clearly constant as functions of b until the weights $w_{ik}(b)$ change as functions of b as a result of altered residual ranks. Therefore, $\Phi_n(b) = A_1 - bA_2$ graphed in Figure 10.1 is piecewise linear.

Following the lead of ordinary least squares in (10.5), the Buckley-James estimator $\widehat{\beta_{0n}}$ of β_0 is the arithmetic mean of the observed partial residuals $Y_i^*(\widehat{\beta_n}) - \widehat{\beta_n}x_i$.

[Figure: A diagram showing a zero-crossing solution to the modified normal equation, with axes labeled y and b, curve $y = \Phi_n(b)$, and point $\widehat{\beta_n}$ marked on the b-axis.]

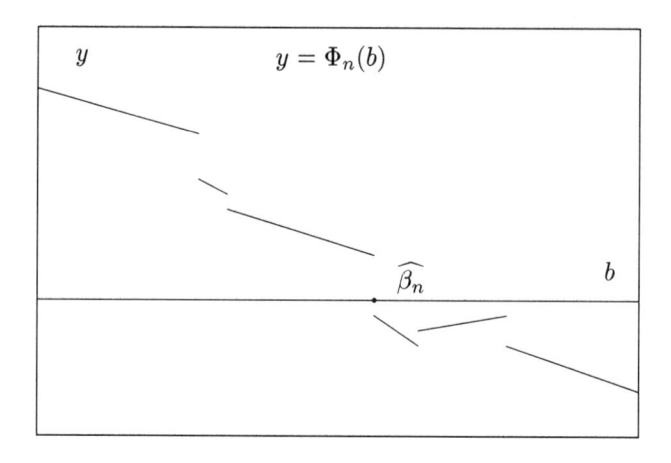

Figure 10.1 *A diagram showing a zero-crossing solution to the modified normal equation $\Phi_n(b) = 0$. The Buckley-James estimate of slope is at the point $b = \widehat{\beta_n}$ where the graph crosses zero. Note that Φ_n is piecewise linear but not necessarily monotone decreasing.*

10.5 Multivariate designs

Adopting vector and matrix notations which are appropriate to multivariate generalisations, we may write the linear model (10.2) with $n \times p$ design matrix $\mathbf{X}$ of full rank p and a first column of 1's, $n \times 1$ residual vector $\boldsymbol{\epsilon}$ with mean $\mathbf{0}$ and $p \times 1$ vector of parameters $\boldsymbol{\beta}$ as

$$\mathbf{Y} = \mathbf{X}\boldsymbol{\beta} + \boldsymbol{\epsilon}. \tag{10.12}$$

We use the term 'scatterplot renovation' to indicate the relocation of censored points in the scatterplot to estimated positions had they been observed without censoring.

The process of scatterplot renovation is then given by an equation which parallels the univariate equation (10.9). The development of this in terms of matrix notation is easier to see if we rewrite (10.9) in terms of summations. Let $\mathbf{b} = (b_0, b_1, \ldots, b_{p-1})^T$ be a vector to hold the parameter values and $\boldsymbol{x}_i^T$ denote the ith row of $\mathbf{X}$. Then

$$Y_i^*(\mathbf{b}) = \boldsymbol{x}_i^T \mathbf{b} + [\epsilon_i(\mathbf{b})\delta_i + \widehat{E_b}(\epsilon_i(\mathbf{b}) \mid \epsilon_i(\mathbf{b}) > c_i(\mathbf{b}))(1 - \delta_i)],$$

and on substituting the weighted linear combination

$$\widehat{E_b}(\epsilon_i(\mathbf{b}) \mid \epsilon_i(\mathbf{b}) > c_i(\mathbf{b})) = \sum_{k=1}^{n} w_{ik}(\mathbf{b}) e_k(\mathbf{b})$$

MULTIVARIATE DESIGNS

for the estimated conditional expected value, we obtain

$$Y_i^*(\mathbf{b}) = \boldsymbol{x}_i^T \mathbf{b} + [\epsilon_i(\mathbf{b})\delta_i + \sum_{k=1}^{n} w_{ik}(\mathbf{b})e_k(\mathbf{b})(1-\delta_i)].$$

Since $w_{ik}(\mathbf{b})(1-\delta_i) = w_{ik}(\mathbf{b})$, we are now in a position to employ matrix notation for the renovation equation. Let

$$\mathbf{Y}^*(\mathbf{b}) = \mathbf{Xb} + \mathcal{W}(\mathbf{b})(\mathbf{Z} - \mathbf{Xb}), \qquad (10.13)$$

where $\mathbf{Z} = (Z_1, Z_2, \ldots, Z_n)^T$, and

$$\mathcal{W}(\mathbf{b}) = \mathbf{diag}(\boldsymbol{\delta}) + [w_{ik}(\mathbf{b})]$$

$$= \begin{pmatrix} \delta_1 & w_{12}(\mathbf{b}) & w_{13}(\mathbf{b}) & \cdots & w_{1n}(\mathbf{b}) \\ 0 & \delta_2 & w_{23}(\mathbf{b}) & \cdots & w_{2n}(\mathbf{b}) \\ \vdots & \vdots & \ddots & \ddots & \vdots \\ 0 & 0 & 0 & \ddots & w_{n-1\ n}(\mathbf{b}) \\ 0 & 0 & 0 & \cdots & \delta_n \end{pmatrix}$$

is a matrix of weights based on slope $\mathbf{b}$ containing the censor indicators $\boldsymbol{\delta}^T = (\delta_1, \delta_2, \delta_3, \ldots, \delta_n)$ on the leading diagonal. The scatterplot renovation is clear in (10.13), where the matrix of weights separates out those censored points which are lifted by the addition of $\mathcal{W}(\mathbf{b})(\mathbf{Z} - \mathbf{Xb})$ and leaves uncensored points untouched.

This gives us a complete description of how to create estimated points $\mathbf{Y}^*(\mathbf{b})$ in the scatterplot based on a given 'slope' $\mathbf{b}$. Following the univariate case of the last section, in order to determine the Buckley-James estimates of $\boldsymbol{\beta}$, we must use the newly created data $\mathbf{Y}^*(\mathbf{b})$ in the least squares normal equation.

The multivariate version of the least squares normal equation from (10.4) is

$$\mathbf{X}^T[\mathbf{Y}^*(\mathbf{b}) - \mathbf{Xb}] = \mathbf{0},$$

which we express in the convenient form

$$\mathbf{b} = (\mathbf{X}^T\mathbf{X})^{-1}\mathbf{X}^T\mathbf{Y}^*(\mathbf{b}). \qquad (10.14)$$

We substitute the $\mathbf{Y}^*(\mathbf{b})$ from (10.13) into this normal equation to obtain

$$\mathbf{b} = (\mathbf{X}^T\mathbf{X})^{-1}\mathbf{X}^T[\mathbf{Xb} + \mathcal{W}(\mathbf{b})(\mathbf{Z} - \mathbf{Xb})].$$

If we set

$$\Phi_n(\mathbf{b}) = (\mathbf{X}^T\mathbf{X})^{-1}\mathbf{X}^T\mathcal{W}(\mathbf{b})(\mathbf{Z} - \mathbf{Xb}), \qquad (10.15)$$

then it is clear that the modified normal equation may be simply expressed as

$$\Phi_n(\mathbf{b}) = \mathbf{0}$$

in a parallel of the univariate case. Therefore, to find parametric estimates of β in (10.12) we must solve $\Phi_n(\mathbf{b}) = \mathbf{0}$. This is achieved through an iterative scheme. Begin with an initial estimate $\mathbf{b} = \mathbf{b}^0$ (possibly using a regression estimate through the uncensored points only). Construct the residuals and matrix of weights based on this slope and hence $\mathbf{Y}^*(\mathbf{b}^0)$. The new values of $\mathbf{b}$ for the next iteration rounds are then

$$\begin{aligned} \mathbf{b}^1 &= (\mathbf{X}^T\mathbf{X})^{-1}\mathbf{X}^T\mathbf{Y}^*(\mathbf{b}^0) \\ \mathbf{b}^2 &= (\mathbf{X}^T\mathbf{X})^{-1}\mathbf{X}^T\mathbf{Y}^*(\mathbf{b}^1) \\ \mathbf{b}^3 &= (\mathbf{X}^T\mathbf{X})^{-1}\mathbf{X}^T\mathbf{Y}^*(\mathbf{b}^2) \\ \ldots &= \ldots \end{aligned}$$

until (a hoped-for) convergence.

As discussed by Buckley and James (1979) and James and Smith (1984), an exact solution to $\Phi_n(\mathbf{b}) = \mathbf{0}$ may not exist. The simplest method (Lin and Wei, 1992) for determining a solution is to compare the vector differences between the left and right sides in the above iterative scheme; that is, find the point where the norm of the vector

$$\Phi_n(\mathbf{b}) = (\mathbf{X}^T\mathbf{X})^{-1}\mathbf{X}^T\mathbf{Y}^*(\mathbf{b}) - \mathbf{b} \tag{10.16}$$

is smallest.

Definition 10.2 $\widehat{\beta_n}$ *is called a* **Buckley-James estimator** *of β in the censored regression model (10.12) if $\mathbf{b} = \widehat{\beta_n}$ minimises the norm of $\Phi_n(\mathbf{b})$.* □

A summary of the steps, called the **Buckley-James method**, involved in finding a Buckley-James estimator $\widehat{\beta_n}$ of β is as follows:

Step 1: Begin with an initial estimate $\mathbf{b} = \mathbf{b}^n$ for $n = 0$;
Step 2: Construct the residuals $\mathbf{Z} - \mathbf{X}\mathbf{b}^n$;
Step 3: Construct the weights matrix $\mathcal{W}(\mathbf{b}^n)$;
Step 4: Construct renovated values $\mathbf{Y}^*(\mathbf{b}^n) = \mathbf{X}\mathbf{b}^n + \mathcal{W}(\mathbf{b}^n)(\mathbf{Z} - \mathbf{X}\mathbf{b}^n)$;
Step 5: Construct next 'slope' $\mathbf{b}^{n+1} = (\mathbf{X}^T\mathbf{X})^{-1}\mathbf{X}^T\mathbf{Y}^*(\mathbf{b}^n)$;
Step 6: Repeat Steps 2 to 5 until $\|\mathbf{b}^{n+1} - (\mathbf{X}^T\mathbf{X})^{-1}\mathbf{X}^T\mathbf{Y}^*(\mathbf{b}^n)\|$ is sufficiently small.

The univariate definition of a Buckley-James estimator is a special case of this definition when, in the univariate case, Φ_n is monotone decreasing in b, because then the zero-crossing corresponds to a point of minimum norm. This monotonicity tends to occur with increases in sample size. (In the exercises is an example where it does not occur for small sample sizes.) If 'solutions' to (10.16) take the form of multiple zero-crossings, or points where Φ_n changes sign, it is customary to take $\widehat{\beta_n}$ as their average. (See, for example, Wu and Zubovic, 1995.)

MULTIVARIATE DESIGNS

Several definitions may now be expressed in terms of the Buckley-James estimator $\widehat{\beta_n}$. Using (10.13), we renovate the scatterplot using $\mathbf{b} = \widehat{\beta_n}$ and, in what follows, suppress dependence of the renovated variables on $\widehat{\beta_n}$ in the notation.

Definition 10.3 *The* **weights matrix**, $\mathcal{W} = \mathcal{W}(\widehat{\beta_n})$, *is the matrix of weights evaluated at the Buckley-James estimator.* □

Definition 10.4 *The* **renovated responses**, $\mathbf{Y}^*$, *are given by the equation* $\mathbf{Y}^* = \mathbf{X}\widehat{\beta_n} + \mathcal{W}(\mathbf{Z} - \mathbf{X}\widehat{\beta_n})$. □

Definition 10.5 *By a* **renovated scatterplot** *we mean (a diagrammatic depiction of) the pair* $(\mathbf{X}, \mathbf{Y}^*)$ *with* $\mathbf{Y}^*$ *as in Definition 10.4.* □

■ **Example 10.1**
Renovating the Stanford Heart Transplant data
When One of the first, and most famous examples in the literature, of a censored regression data set is the so-called Stanford Heart Transplant data (Example 5.2), first detailed by Crowley and Hu (1977). We show this scatterplot in renovated form in Figure 10.2 for a single 'Age' covariate. Although many covariates were considered in the first analyses of the Stanford Heart Transplant Data, we show here a scatterplot of 'Time since heart transplant' in 'log days' (base 10) against 'Age' in years. The vertical lines in Figure 10.2 show the extent of the renovation of censored points from base to tip of a given line. These data are somewhat diffuse, with Age being just significant (Buckley and James, 1979). These data will be discussed further in Chapter 11 from the point of view of regression diagnostics. □

Properties of the weights derived from considerations of the self-consistency of the product-limit estimator (Efron, 1967; Buckley and James, 1979; Smith and Zhang, 1995) and direct inspection provide the following properties of the weights matrix:

Theorem 10.2 *The weights matrix $\mathcal{W}$ is upper-triangular and satisfies:*

1. $\mathcal{W}^2 = \mathcal{W}$ and $(\mathbf{I} - \mathcal{W})^2 = \mathbf{I} - \mathcal{W}$ *(idempotence);*
2. $\mathcal{W}\mathbf{1} = \mathbf{1}$ *where* $\mathbf{1}$ *is an* $n \times 1$ *vector of 1's (row sums);*
3. $\mathbf{1}^T\mathcal{W} = n\mathbf{v}^T$, *where* $\mathbf{v} = (v_1, v_2, \ldots v_n)^T$ *(column sums);*
4. $\mathcal{W} = \mathbf{I}$, *the* $n \times n$ *identity matrix, in the absence of any censoring.*

□

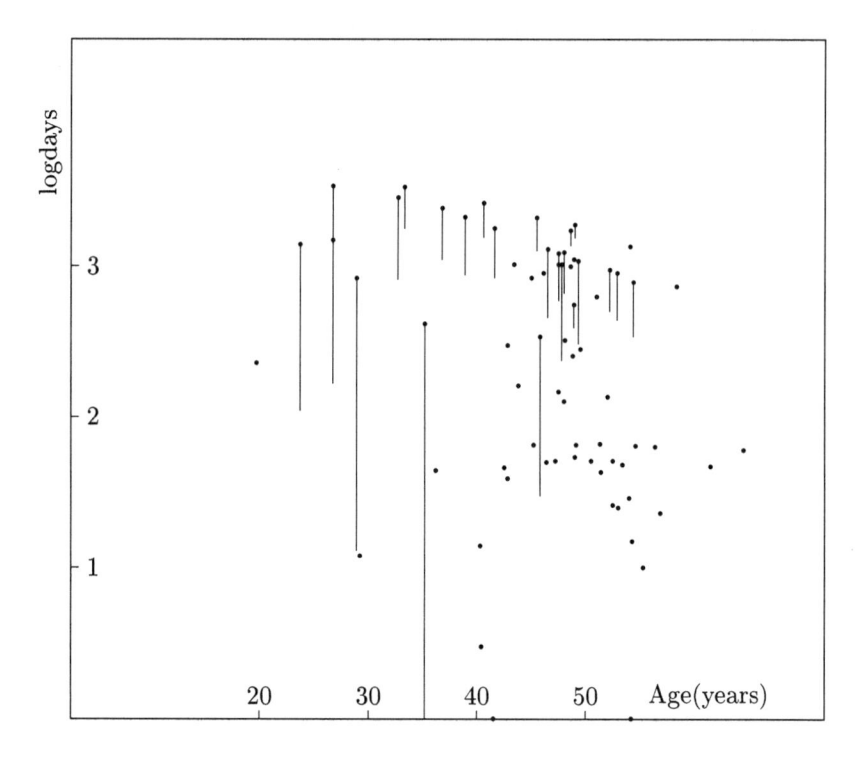

Figure 10.2 *A renovated scatterplot for the first 69 patients in the Stanford Heart Transplant Program. Vertical lines trace the extent of the renovation of patients with right-censored lifetimes.*

10.6 Properties of Buckley-James estimators

It is certainly true that properties of the Buckley-James estimators have proved difficult to establish. In early work on the univariate case, James and Smith (1984) established weak consistency of the iterative solution $\widehat{\beta_n}$ to

$$\Phi_n(b) = \frac{\sum_{i=1}^n (x_i - \bar{x}) Y_i^*(b)}{\sum_{i=1}^n (x_i - \bar{x})^2} - b = 0.$$

They established under certain regularity conditions that for and $\delta > 0$, with probability tending to 1 as $n \to \infty$, there exists a point $b = \widehat{\beta_n}$ in $(\beta - \delta, \beta + \delta)$ at which $\Phi_n(b)$ changes sign. This guaranteed the existence of a Buckley-James estimator near the true parameter with probability tending to 1.

One of the most important conditions necessary for this result is that, if we define $\mathcal{N}_b(u)$ to be the expected number of censored and uncensored values $e_i(b)$ exceeding u, then

$$\mathcal{N}_b(u) \to \infty \text{ as } n \to \infty \text{ for all } u \in (-\infty, U_b], \tag{10.17}$$

where U_b lies at the top of the support of the residual distribution. (10.17) declares that potentially large numbers of both censored and uncensored residuals, particularly near the top of the distribution are needed to ensure the uniform consistency of the product-limit estimator, in its role as estimator of the 'survival function' of the residuals. Instabilities in the estimation of S for values of u where $S(u)$ is near 0, cloud the ability of

$$\widehat{E_b}(\epsilon_i(b) \mid \epsilon_i(b) > c_i(b)) = \sum_{k=1}^{n} w_{ik}(b) e_k(b)$$

to successfully estimate

$$E(\epsilon_i(b) \mid \epsilon_i(b) > c_i(b)) = c_i(b) + \int_{c_i(b)}^{U_b} \frac{S_{n,b}(u)}{S_{n,b}(c_i(b))} du.$$

Here, $S_{n,b}(u)$ is the survival function of the $\epsilon_i(b) = y_i - bx_i$ and the upper bound, U_b, is given by $U_b = \sup\{u : S_{n,b}(u) > 0\} < \infty$.

Certainly, at the true parameter $b = \beta = \beta_1$, the following result holds (James and Smith, 1984):

Theorem 10.3 *If $\mathcal{N}_\beta(u) \to \infty$ as $n \to \infty$ for all $u < T = U_\beta$,*

$$\sup_{u \in (-\infty, T)} |\widehat{E_\beta}(\epsilon_i \mid \epsilon_i > c) - E_\beta(\epsilon_i \mid \epsilon_i > c)| \to_p 0,$$

where $\widehat{E_\beta}(\epsilon_i \mid \epsilon_i > c)$ is the estimator of $E_\beta(\epsilon_i \mid \epsilon_i > c)$ based on the product-limit estimator $\widehat{S_{n,\beta}}$ of the survival function $S_{n,\beta}$. □

(Recall that random variables V_n **converge in probability** to v, written $V_n \to_p v$, if for all $\epsilon > 0$, $\lim_{n\to\infty} P(|V_n - v| > \epsilon) = 0$.)

Note that at values of b away from the true parameter, the exact form of $S_{n,b}(u)$ is a complex structure. This is because although ϵ_i may be independent and identically distributed residuals in the original model (10.2), the residuals

$$\epsilon_i(b) = \epsilon_i + (\beta - b)x_i$$

based on a slope b different from the true parameter are no longer independent of the censoring variables

$$c_i(b) = c_i + (\beta - b)x_i.$$

The distribution of the x_i is involved in both expressions. Lai and Ying (1991) were able to show that

$$S_{n,b}(t) = \exp\left\{ -\int_{-\infty}^{t} \frac{\int_{-\infty}^{\infty} f(u + (b-\beta)x)\overline{G_n}(x, u + (b-\beta)x) dH(x)}{\int_{-\infty}^{\infty} S(u + (b-\beta)x)\overline{G_n}(x, u + (b-\beta)x) dH(x)} du \right\},$$
(10.18)

where the x_i are regarded as independent and identically distributed with a common distribution function H,

$$G_i(x, s) = P(t_i - \beta x_i \geq s \mid x_i = x),$$

$\overline{G}_n = n^{-1} \sum_{i=1}^{n} G_i$ and $f = -S'$. See Lai and Ying (1991) for the derivation of (10.18) as the limiting value of their modified product-limit estimator $\widehat{S_{n,b}}(t)$.

If, however, the censoring is reasonably uniform along the line so that in (10.18), $\overline{G}_n(x, u + (b-\beta)x) \approx G(u)$ for b near β, then $S_{n,b}(t)$ simplifies to the formula

$$S_b(t) = \int_{-\infty}^{\infty} S(t + (b-\beta)x) dH(x)$$

which James and Smith (1984) proposed as the survival function for residuals based on slope b. Note that, at the true parameter, $S_{n,\beta}(u) = S_\beta(u) = S(u)$.

Distributional properties of $\widehat{\beta_n}$ are succinctly outlined in Lai and Ying (1991), albeit for a slightly different censorship model and a smoothed version of the product-limit estimator. The basic distributional results, subject to regularity conditions, are given in Theorem 10.4.

Theorem 10.4 *For $r = 0, 1, 2$, define $\Gamma_r(t)$ by*

$$\frac{\sum_{i=1}^{n}(x_i - \bar{x})^r G_i(x_i, t)}{n} \to_p \Gamma_r(t).$$

Let $\tau = \sup\{t : S(t)\Gamma_0(t) > 0\}$. Then under certain regularity conditions, $\sqrt{n}(\widehat{\beta_n} - \beta)$ has a limiting asymptotic normal distribution with mean 0 and variance

$$\frac{\int_{-\infty}^{\tau}\left\{\Gamma_2(t) - \frac{\Gamma_1^2(t)}{\Gamma_0(t)}\right\} \frac{\lambda(t)}{S(t)} \left\{\int_t^\tau S(u)du\right\}^2 dt}{\left[\int_{-\infty}^{\tau}\left\{\Gamma_2(t) - \frac{\Gamma_1^2(t)}{\Gamma_0(t)}\right\} \lambda'(t) \left\{\int_t^\tau S(u)du\right\} dt\right]^2}$$

where $\lambda(u)$ is the hazard function for the residuals ϵ_i having derivative $\lambda'(u)$.

□

The results of Theorem 10.4 clearly show that the asymptotic variance of the Buckley-James estimator $\widehat{\beta_n}$ depends on the hazard function (and by implication, the density) of the residuals. This creates difficulties for estimation purposes; $Var(\widehat{\beta_n})$ remains difficult to estimate. An obvious suggestion, for large sample sizes, is to use classical lifetable estimators of the hazard function. These are outlined in, for example, Smith (1986) and successfully used by Hillis (1993 and 1994). Note that the asymptotic variance formula in Theorem 10.4 simplifies considerably under the assumption that censoring is even along the true regression line, since then we may write $G_i(x_i, t) = G(t)$, say, so that $\Gamma_1(t) = 0$. Under these conditions, Smith (1986) uses an asymptotic variance formula for $\widehat{\beta_n}$ of the form

$$Var\left[\sqrt{\sum_{i=1}^{n}(x_i - \bar{x})^2}\,\widehat{\beta_n}\right] \sim \frac{\Sigma_n(\beta)}{\psi_n(\beta)^2},$$

where
$$\Sigma_n(\beta) = \frac{\sum_{i=1}^n (x_i - \bar{x})^2 \sigma_i^2}{\sum_{i=1}^n (x_i - \bar{x})^2}, \qquad (10.19)$$

for
$$\sigma_i^2 = \sigma^2 - Var[\epsilon_i | \epsilon_i > c_i] S(c_i)$$

and
$$\psi_n(\beta) = \frac{\sum_{i=1}^n (x_i - \bar{x})^2 S(c_i)[1 - \lambda(c_i) \int_{c_i}^T S(u) du]}{\sum_{i=1}^n (x_i - \bar{x})^2} - 1. \qquad (10.20)$$

It must be emphasised that this asymptotic variance is only valid under the restrictive case where censoring is even along the line for values of b near β. How can we use this result to then estimate $Var(\widehat{\beta_n})$? We use $\widehat{\beta_n}$, the product-limit estimator based on $z_i - \widehat{\beta_n} x_i$ and appropriate hazard estimates such as detailed in Smith (1986) and Hillis (1993). Since not all the censor times are known, we use the standard procedure of estimating the unknown terms $S(c_i)$ in (10.20) by $1 - \delta_i$. This procedure was followed by Buckley and James (1979), where they also provide a further heuristic variance estimator.

Further, since it may be shown (Smith, 1988) for uniform censoring along the line, that at the true parameter, $\Sigma_n(\beta)$ at (10.19) approximates the variance $Var[\sqrt{\sum_{i=1}^n (x_i - \bar{x})^2} \Phi_n(\beta)]$, then a 95% confidence interval for β may be constructed without the need for evaluating $Var(\widehat{\beta_n})$, by 'inverting the test statistic' created by appropriately normalising $\Phi_n(\beta)$. That is, since asymptotically,

$$\frac{\sqrt{\sum_{i=1}^n (x_i - \bar{x})^2} \Phi_n(\beta)}{\sqrt{\Sigma_n(\beta)}} \approx N(0, 1),$$

an approximate confidence region for β is found, for n large, by including in the region those values of β which, when substituted into the test statistic

$$\frac{\sqrt{\sum_{i=1}^n (x_i - \bar{x})^2} \Phi_n(\beta)}{\sqrt{\Sigma_n(\beta)}},$$

give output in the range $(-2, 2)$. More specifically, a range $(-1.96, 1.96)$ is appropriate for a 95% confidence interval. However, a lack of monotonicity in $\Phi_n(\beta)$, may create a region of values for the 'confidence interval' rather than an interval of values in the usual manner.

10.7 Renovation and least squares

In this section, we return to multivariate notation and concentrate on the residuals created when a Buckley-James 'line' is fitted to a renovated scatterplot.

Consider the so-called **renovated residuals**, ϵ^*, available in the renovated scatterplot:
$$\epsilon^* = \mathbf{Y}^* - \mathbf{X} \widehat{\beta_n}$$

Cross-multiplying the renovation equation $\mathbf{Y}^* = \mathbf{X}\widehat{\beta_n} + \mathcal{W}(\mathbf{Z} - \mathbf{X}\widehat{\beta_n})$ by $\mathcal{W}$ shows that $\mathcal{W}\mathbf{Z} = \mathcal{W}\mathbf{Y}^*$, so that we may write the renovation equation in the form

$$\mathbf{Y}^* = \mathbf{X}\widehat{\beta_n} + \mathcal{W}(\mathbf{Y}^* - \mathbf{X}\widehat{\beta_n}). \tag{10.21}$$

Subtracting $\mathbf{X}\widehat{\beta_n}$ from each side of this equation shows that the renovated residuals are constant under $\mathcal{W}$:

$$\epsilon^* = \mathcal{W}\epsilon^*$$

Therefore, the renovated residuals produce the same renovated scatterplot (X, Y^*) as location-shifted residuals $\epsilon^* + c\mathbf{1}$ where c is a scalar. For this reason the renovation process may be completed prior to intercept estimation.

Following the work of Smith and Zhang (1995), we examine the relationship between scatterplot renovation and least squares estimation. We define the **least squares estimator** $\mathbf{b} = \widehat{\beta_{LS}}$ of β by the equation

$$\widehat{\beta_{LS}} = (\mathbf{X}^T\mathbf{X})^{-1}\mathbf{X}^T\mathbf{Y}^*,$$

where $\mathbf{Y}^* = \mathbf{X}\widehat{\beta_n} + \mathcal{W}(\mathbf{Z} - \mathbf{X}\widehat{\beta_n})$. The relationship between $\widehat{\beta_{LS}}$ and $\widehat{\beta_n}$ is given by the following theorem.

Theorem 10.5 *The Buckley-James estimator of β and the least squares estimator of β in the renovated scatterplot coincide precisely when $\Phi_n(\widehat{\beta_n}) = 0$ and $\widehat{\beta_n}$ is an exact zero-crossing.*

Proof: Cross-multiplying each side of (10.4) by $(\mathbf{X}^T\mathbf{X})^{-1}\mathbf{X}^T$, it follows immediately that

$$\widehat{\beta_{LS}} = \widehat{\beta_n} + \Phi_n(\widehat{\beta_n}).$$

This clearly shows that the least squares estimator and the Buckley-James estimator coincide when the zero-crossing is exact. □

The following theorem shows that $\widehat{\beta_n}$ inherits a least-squares-type structure from $\widehat{\beta_{LS}}$ even when the zero-crossing is not exact.

Theorem 10.6 *Suppose that the Buckley-James estimator $\mathbf{b} = \widehat{\beta_n}$ is an exact solution of (10.15), so that $\Phi_n(\widehat{\beta_n}) = 0$. Then $\widehat{\beta_n} = (\mathbf{X}^T\mathcal{W}\mathbf{X})^{-1}\mathbf{X}^T\mathcal{W}\mathbf{Y}^*$ is the least squares estimator of β in the renovated scatterplot $(\mathbf{X}, \mathbf{Y}^*)$.*

If $\Phi_n(\widehat{\beta_n}) \neq 0$, then $\widehat{\beta_n} = (\mathbf{X}^T(\mathbf{I} - \mathcal{W})\mathbf{X})^{-1}\mathbf{X}^T(\mathbf{I} - \mathcal{W})\mathbf{Y}^*$, which is not the least squares estimator of β.

Proof: Suppose initially that $\Phi_n(\widehat{\beta_n}) = 0$. Then, expanding (10.15) it follows that $\mathbf{X}^T\mathcal{W}\mathbf{Z} - \mathbf{X}^T\mathcal{W}\mathbf{X}\mathbf{b} = 0$. Therefore, if $\mathbf{X}^T\mathcal{W}\mathbf{X}$ is invertible, then $\widehat{\beta_n} = (\mathbf{X}^T\mathcal{W}\mathbf{X})^{-1}\mathbf{X}^T\mathcal{W}\mathbf{Z}$. Since the weights matrix is idempotent, $\mathcal{W}^2 = \mathcal{W}$, so that multiplication by $\mathcal{W}$ on each side of (10.13) shows that the renovated data and the original data are constant under $\mathcal{W}$. That is, $\mathcal{W}\mathbf{Z} = \mathcal{W}\mathbf{Y}^*$. The solution $\widehat{\beta_n}$ may then be written as

$$\widehat{\beta_n} = (\mathbf{X}^T\mathcal{W}\mathbf{X})^{-1}\mathbf{X}^T\mathcal{W}\mathbf{Y}^*.$$

By construction, $\mathbf{b} = \widehat{\beta_n}$ specifies the least squares estimator in the renovated scatterplot $(\mathbf{X}, \mathbf{Y}^*)$. Finally, if $\Phi_n(\widehat{\beta_n}) \neq 0$, writing $\widehat{\beta_{LS}} = \widehat{\beta_n} + \Phi_n(\widehat{\beta_n})$ in the form

$$(\mathbf{X}^T\mathbf{X})^{-1}\mathbf{X}^T\mathbf{Y}^* = \widehat{\beta_n} + (\mathbf{X}^T\mathbf{X})^{-1}\mathbf{X}^T\mathcal{W}(\mathbf{Y}^* - \mathbf{X}\widehat{\beta_n}),$$

it follows through cross-multiplying by $(\mathbf{X}^T\mathbf{X})$ and collecting terms in $\mathbf{Y}^*$ and $\widehat{\beta_n}$ that

$$\mathbf{X}^T(\mathbf{I} - \mathcal{W})\mathbf{Y}^* = \mathbf{X}^T(\mathbf{I} - \mathcal{W})\mathbf{X}\widehat{\beta_n}.$$

Therefore, the final statement of the theorem follows: when $\mathbf{X}^T(\mathbf{I} - \mathcal{W})\mathbf{X}$ is invertible, $\widehat{\beta_n} = (\mathbf{X}^T(\mathbf{I} - \mathcal{W})\mathbf{X})^{-1}\mathbf{X}^T(\mathbf{I} - \mathcal{W})\mathbf{Y}^*$. □

The equation $\mathbf{X}^T(\mathbf{I} - \mathcal{W})\mathbf{Y}^* = \mathbf{X}^T(\mathbf{I} - \mathcal{W})\mathbf{X}\widehat{\beta_n}$ shows the relationship between the various forms of solution in the renovated scatterplot. In fact, simple rearrangement gives

$$\mathbf{X}^T\mathbf{Y}^* - \mathbf{X}^T\mathbf{X}\widehat{\beta_n} = \mathbf{X}^T\mathcal{W}\mathbf{Y}^* - \mathbf{X}^T\mathcal{W}\mathbf{X}\widehat{\beta_n}.$$

The right-side of this last equation is 0 for an exact zero-crossing, thus providing the same solution for the least squares slope on the left-side.

This theorem demonstrates that the Buckley-James estimators have a least-squares-type structure $\widehat{\beta_n} = (\mathbf{X}^T\mathbf{A}\mathbf{X})^{-1}\mathbf{X}^T\mathbf{A}\mathbf{Y}^*$, where $\mathbf{A}$ is idempotent, regardless of whether the zero-crossing is 'exact' or not. This means that the usual least squares regression theory for points of influence, points of high leverage and regression diagnostics may be applied to Buckley-James estimators by paralleling the properties enjoyed by least squares estimators in the renovated scatterplot.

Consider the two types of residuals available in the renovated scatterplot: $\epsilon^* = \mathbf{Y}^* - \mathbf{X}\widehat{\beta_n}$ or **renovated residuals**, and $\epsilon_{LS} = \mathbf{Y}^* - \mathbf{X}\widehat{\beta_{LS}}$ or **least squares residuals**. The residual transformation

$$\epsilon^* = \mathcal{W}\epsilon^*$$

characterises the Buckley-James method. It is different, both numerically and in spirit, from regression methods such as **weighted least squares**, where both the response and covariate are linearly transformed. That is, the least squares estimator based on transformed covariate $\mathcal{W}\mathbf{X}$ and transformed response $\mathcal{W}\mathbf{Z}$ is

$$\mathbf{b} = (\mathbf{X}^T\mathcal{W}^T\mathcal{W}\mathbf{X})^{-1}\mathbf{X}^T\mathcal{W}^T\mathcal{W}\mathbf{Z}^*,$$

which is essentially different from $\widehat{\beta_n} = (\mathbf{X}^T\mathcal{W}\mathbf{X})^{-1}\mathbf{X}^T\mathcal{W}\mathbf{Y}^*$ since $\mathcal{W}$ is not symmetric. Indeed, $\mathbf{b} = (\mathbf{X}^T\mathcal{W}^T\mathcal{W}\mathbf{X})^{-1}\mathbf{X}^T\mathcal{W}^T\mathcal{W}\mathbf{Z}^*$ gives the least squares slope through the scatterplot $(\mathcal{W}\mathbf{X}, \mathcal{W}\mathbf{Y}^*)$.

Note that $\widehat{\beta_{LS}} = \widehat{\beta_n} + \Phi_n(\widehat{\beta_n})$ may now be rewritten as

$$\widehat{\beta_{LS}} = \widehat{\beta_n} + (\mathbf{X}^T\mathbf{X})^{-1}\mathbf{X}^T\epsilon^*.$$

This allows the comparison of fitted values $\mathbf{X}\widehat{\beta_n}$ and $\mathbf{X}\widehat{\beta_{LS}}$ from the renovation and least squares procedures be compared: by (10.21), and in terms of the usual regression **hat matrix** $\mathbf{H} = \mathbf{X}(\mathbf{X}^T\mathbf{X})^{-1}\mathbf{X}^T$,

$$\mathbf{X}\widehat{\boldsymbol{\beta}_{LS}} = \mathbf{X}\widehat{\boldsymbol{\beta}_n} + \mathbf{H}\boldsymbol{\epsilon}^*;$$

and for a comparison of the residuals,

$$\boldsymbol{\epsilon}_{LS} = (\mathbf{I} - \mathbf{H})\boldsymbol{\epsilon}^*.$$

The least squares and renovated slope estimates coincide when the hat matrix annihilates the renovated residuals; that is, when $\mathbf{H}\boldsymbol{\epsilon}^* = \mathbf{0}$.

10.8 Interval-censored responses

In this section we apply the Buckley-James censored regression methods described in the last section to single-sample and multi-sample interval-censored responses as described in Smith (1996). The renovation effect corresponds to the vertical movement of points in a scatterplot in the single-sample case, or a move to the right of points along the response axis when comparing several samples. In all cases, each point is moved to an estimated position within its own censoring interval. Such an estimation process encourages the use of the self-consistent product-limit estimator of the survival function (Chapter 6) because this estimator is appropriate for right-censored data. We demonstrate how the same renovation methodology then applies to what has been traditionally considered structurally different left-, right- and interval-censored data.

We begin by describing how interval-censored data may appear in the regression context. Suppose that the outcomes $y_1, y_2, \ldots, y_n$ of a positive-valued response variable Y are subject to a type of **interval-censoring** capable of wide application to industrial inspection programmes and medical follow-up studies. For each i, it is known only that the ith response y_i lies in a **censoring interval** I_i having known specified endpoints. In many applications such endpoints are obtained from a list $0 \leq t_0 < t_1 < t_2 < \ldots < t_m \leq \infty$ of programmed inspection times; for example, monthly medical checks, hourly industrial inspections. Adjacent finite inspection times may be chosen to differ by one time unit; however, the ordinality of the inspection times and their independence of the response being measured, is all that the regression methodology requires. The t_i's need not be integers.

Accordingly, for each i there are endpoints t_{i_1} and t_{i_2} with indices $i_1 \leq i_2$ depending on i, so that $I_i = [t_{i_1}, t_{i_2}]$, where

$i_2 = i_1$ for *observed failure at* t_{i_1};
$i_2 > i_1$ for *interval-censoring* including:
$t_{i_2} = \infty$ for *right-censoring*;
$t_{i_1} = 0$ for *left-censoring*.

Note that observed failure occurs when an interval has coincident endpoints. (Peto, 1973; Turnbull, 1976.)

The observed data are recorded in the form

$$(x_1, I_1), (x_2, I_2), \ldots, (x_n, I_n),$$

where x_i is the covariate associated the interval-censored response. In the multiple-sample problem, x_i denotes group membership and n represents the combined sample size. Generally, we identify I_i with its corresponding left-hand endpoint z_i, while retaining knowledge of interval width and consequent censoring status. Thus, a censor interval $I_i = [t_{i_1}, t_{i_2}]$ provides an observed data point $z_i = t_{i_1}$, an interval width $w_i = t_{i_2} - t_{i_1}$, as well as an indicator of censoring: $\delta_i = 1$ if there is an observed failure at $z_i = t_{i_1}$ so that $w_i = 0$; $\delta_i = 0$ when there is left-, right-, interval-censoring present for which $w_i > 0$. For examples of such interval-censored data sets, see Finkelstein and Wolfe (1985) or Collett (1994, page 239).

Note that, completely equivalently, the observed data are

$$(x_1, z_1, w_1), (x_2, z_2, w_2), \ldots, (x_n, z_n, w_n),$$

and the censor indicators are given by

$$\delta_i = \begin{cases} 1 & \text{if } w_i = 0; \\ 0 & \text{if } w_i > 0. \end{cases}$$

Our aim is to re-position each data point z_i by moving it conditionally to the right along the response axis — to re-place it inside its own censoring interval I_i thereby providing a view of the data in the absence of censoring.

	t_0	t_1	t_2	t_3	t_4	t_5	t_6	t_7	t_8	t_9	t_{10}
$Unit\ 1$	•	•	•	•	•	•	•	•	•	•	•
$Unit\ 2$	•	•	•	•	•	·	·	·	○		
$Unit\ 3$	•	•	•	•	•	•	•	○			
$Unit\ 4$	·	·	·	○							

$I_1 = \{t_{10}\}$ $z_1 = t_{10}$ $\delta_1 = 1$
$I_2 = [t_4, t_8]$ $z_2 = t_4$ $\delta_2 = 0$
$I_3 = [t_7, \infty]$ $z_3 = t_7$ $\delta_3 = 0$
$I_4 = [0, t_3]$ $z_4 = 0$ $\delta_4 = 0$

Figure 10.3 *Four units providing censored data: Units 1–4 respectively show observed failure, interval-censoring, right-censoring, and left-censoring.*

The interval-censoring notation applying to different types of censoring is illustrated in Figure 10.3 which gives a time line for four units whose lifetimes we seek to record. The type of censoring experienced by the units may be represented by a sequence of dots recorded at the inspection times. '•' indicates that the unit was operational at inspection; 'o' indicates that the unit had failed at inspection time; '·' indicates that the unit was unobservable at inspection time.

This interval-censoring notation accommodates *double-censoring*, where each observation y_i is either left-censored, or right-censored, or observed exactly in a window $[l_i, u_i]$ on the response axis between two inspection times, giving observed data $z_i = \max\{\min\{y_i, u_i\}, l_i\}$. (Turnbull, 1974; Chang and Yang, 1987.) In this case, $I_i = [0, l_i]$ or $[u_i, \infty]$ for intervals in which censoring occurs; exact responses are singleton sets inside $[l_i, u_i]$.

In the most general setting, we wish to analyse responses which are all either observed, left-, right-, or interval-censored. However, the renovation process *depends* on the presence (of at least some, and advisedly, half) of responses observed exactly, whereas the methods of Peto (1973) and Turnbull (1974, 1976) for the estimation of the survival function may be applied to data where every data point is censored to an interval.

In the Buckley-James method we fit the model (10.12) which we restate as

$$\mathbf{Y} = \mathbf{X}\boldsymbol{\beta} + \boldsymbol{\epsilon}.$$

As before, censored responses in the scatterplot are replaced by their estimated conditional expected responses using a weighted linear combination of observed residuals $\mathbf{E}(\mathbf{b}) = (e_1(\mathbf{b}), e_2(\mathbf{b}), \ldots, e_n(\mathbf{b}))^T = \mathbf{Z} - \mathbf{X}\mathbf{b}$ from a fitted line of 'slope' $\mathbf{b}$ through the observed left-hand endpoints $\mathbf{Z} = (Z_1, Z_2, \ldots, Z_n)^T$. Let $\widehat{F_b}$ denote the product-limit estimator based on the observed residual vector $\mathbf{E}(\mathbf{b})$. If we create residual intervals $J_i(\mathbf{b})$ corresponding to the response intervals I_i by subtracting the fitted values $\mathbf{X}\mathbf{b}$, the weights

$$w_{ik}(\mathbf{b}) = \begin{cases} \dfrac{d\widehat{F_b}(e_k(\mathbf{b}))\delta_k(1-\delta_i)}{\displaystyle\int_{J_i(\mathbf{b})} d\widehat{F_b}(u)} & \text{if } e_k(\mathbf{b}) \in J_i(\mathbf{b}) \\ 0 & \text{otherwise} \end{cases} \quad (10.22)$$

are a generalisation, appropriate to interval-censoring, of the similar fractions given in (10.10). For a residual interval $[l, u]$, if the true regression parameters were known, the weights are designed so as to approximate

$$\frac{dF(r)}{S(l) - S(u)}$$

in the equation

$$E[R_i | l < R_i < u] = \int_l^u r \frac{dF(r)}{S(l) - S(u)}.$$

[Note that $u = \infty$ for right-censored observations; $l = -\infty$ for left-censored observations.] This means that when $\mathbf{b}$ is near $\boldsymbol{\beta}$, the conditional expected residual size $E[R_i | R_i \in J_i(\boldsymbol{\beta})]$ is estimated by a linear combination

$$\sum_{e_k(\mathbf{b}) \in J_i(\mathbf{b})} e_k(\mathbf{b}) w_{ik}(\mathbf{b}).$$

In a multivariate setting, such as for comparing more than two samples, the Buckley-James method consists of determining an iterative solution $\mathbf{b} = \widehat{\boldsymbol{\beta}_n}$ to the equation

$$(\mathbf{X}^T\mathbf{X})^{-1}\mathbf{X}^T\mathbf{Y}^*(\mathbf{b}) - \mathbf{b} = 0 \quad (10.23)$$

through the renovated responses $\mathbf{Y}^*(\mathbf{b}) = \mathbf{X}\mathbf{b} + \mathcal{W}(\mathbf{b})(\mathbf{Z} - \mathbf{X}\mathbf{b})$, where

$$\mathcal{W}(\mathbf{b}) = diag(\boldsymbol{\delta}) + [w_{ik}(\mathbf{b})]$$

is the upper triangular **weights matrix** containing the censor indicators on the main diagonal. A solution to (10.23) is reached iteratively when the norm of the left-side is minimum (James and Smith, 1984; Lin and Wei, 1992a).

Once the Buckley-James solution $\mathbf{b} = \widehat{\beta_n}$ has been found, the response data comprising the left-hand ends of the censoring intervals may be 'renovated' to construct new responses $\mathbf{Y}^*$, where

$$\mathbf{Y}^* = \mathbf{Y}^*(\widehat{\beta_n}) = \mathbf{X}\widehat{\beta_n} + \mathcal{W}(\mathbf{Z} - \mathbf{X}\widehat{\beta_n}) \tag{10.24}$$

and $\mathcal{W} = \mathcal{W}(\widehat{\beta_n})$.

The renovation process described described in the previous section may be constructed in the interval-censoring framework. The weights matrix $\mathcal{W}$ now has a specialised structure that may be seen as a generalisation of that applying to right-censored responses. Even so, it is easily seen to satisfy: $\mathcal{W}^2 = \mathcal{W}$; $(\mathbf{I} - \mathcal{W})^2 = \mathbf{I} - \mathcal{W}$; $\mathcal{W}\mathbf{1} = \mathbf{1}$ where $\mathbf{1}$ is an $n \times 1$ vector of 1's. Then, since $\mathcal{W}\mathbf{Z} = \mathcal{W}\mathbf{Y}^*$, we may write (10.24) as

$$\mathbf{Y}^* = \mathbf{X}\widehat{\beta_n} + \mathcal{W}(\mathbf{Y}^* - \mathbf{X}\widehat{\beta_n}). \tag{10.25}$$

As before, $\widehat{\beta_n}$ has the least-squares-type structure

$$\widehat{\beta_n} = (\mathbf{X}^T \mathcal{W} \mathbf{X})^{-1} \mathbf{X}^T \mathcal{W} \mathbf{Y}^* \tag{10.26}$$

if the zero-crossing in (10.23) is exact. Therefore we are returned to the least squares methodology of the previous section.

The most important feature of the renovation process applied to interval-censored data is that the responses may now be ranked so that comparisons can be made in diagnostic plotting. For example, when comparing two interval-censored samples, standard boxplot comparisons, such as those described by Smith (1996) for comparing right-censored samples, may then take place on the renovated data.

▬ Example 10.2
Renovating trial interval-censored data

The renovation process for artificial interval-censored data is demonstrated with given response intervals I_i, for $i = 1, 2, 3, \ldots, 40$ which represent interval-censored observations of a survival response variable in units with data listed in the table. For example, a left-censored observation of 15 units is represented by the interval $[0, 15]$; a right-censored observation of 8 units is represented by the interval $[8, \infty]$; and an observation which is interval-censored on the left by 15 units and the right by 8 units is represented by [8, 15]; an observed death at 6 units is represented by {6}.

	Original trial data (I)		Renovated trial data (y^*)
Sample 0	Sample 1	Sample 0	Sample 1
[0, 15]	[0, 17]	12.09	13.22
[0, 22]	[0, 28]	15.02	21.57
[0, 30]	[0, 25]	18.98	19.94
{6}	[0, 40]	6.00	26.20
[8, ∞]	[17, 30]	21.10	23.38
[8, 15]	{18}	12.69	18.00
[9, ∞]	{18}	21.09	18.00
{12}	[18, ∞]	12.00	29.11
[12, 20]	{20}	15.92	20.00
[12, 25]	{20}	18.07	20.00
{13}	{21}	13.00	21.00
[14, ∞]	{25}	24.35	25.00
{15}	{26}	15.00	26.00
{18}	[26, 33]	18.00	29.69
{20}	[27, 35]	20.00	31.80
{22}	{30}	22.00	30.00
[25, ∞]	[31, ∞]	28.93	35.55
[25, 30]	{32}	26.50	32.00
{26}	[32, ∞]	26.00	36.15
{27}	{41}	27.00	41.00

We show two samples, Sample 0 and Sample 1 each of size 20, such that

$$x_i = \begin{cases} 0 & \text{if } I_i \text{ is from Sample 0} \\ 1 & \text{if } I_i \text{ is from Sample 1.} \end{cases}$$

The original interval-censored data, expressed as intervals, and the renovated data, expressed as two data sets, are as in the table.

The Buckley-James procedure applied to the two interval-censored samples produced two 'close-together' slope estimates of 7.17 and 7.27. These values were averaged for the renovation process, giving a Buckley-James estimator $\widehat{\beta_n} = (18.67, 7.22)^T$. The observed values of $\mathbf{Y}^* = \mathbf{X}\widehat{\beta_n} + \mathcal{W}(\mathbf{Y}^* - \mathbf{X}\widehat{\beta_n})$ comprise the renovated samples listed in the table.

The samples, in their original form as collections of intervals, are difficult to compare since overlapping intervals do not easily permit an ordering. However, under the assumption of an underlying linear model, the renovation process produces a combined ranked 'data set' which may be appropriately displayed by rank procedures such as boxplots. This facilitates direct visual comparison of the data sets concerned. □

10.9 Exercises

10.1. Using the simple linear model

$$Y_i = \beta_0 + \beta_1 x_i + \epsilon_i; \quad \epsilon_i \sim F; \quad E(\epsilon_i) = 0; \quad Var(\epsilon_i) = \sigma^2$$

with right-censoring $Z_i = \min\{Y_i, t_i\}$ of the response variable, where the t_i are fixed censor constants with

$$\delta_i = \begin{cases} 1 & \text{if } y_i < t_i \\ 0 & \text{if } y_i \geq t_i, \end{cases}$$

show that the random quantities

$$\epsilon_i \delta_i + E(\epsilon_i | \epsilon_i > c_i)(1 - \delta_i)$$

in terms of $c_i = t_i - \beta_0 - \beta_1 x_i$ have mean 0.

10.2. Show using (10.11) that

$$\Phi_n(b) = \frac{\sum_{i=1}^n (x_i - \bar{x})[e_i \delta_i + \sum_{k=1}^n w_{ik}(b) e_k (1 - \delta_i)]}{\sum_{i=1}^n (x_i - \bar{x})^2}.$$

Use this result to show that $\Phi_n(b)$ is piecewise linear.

Show further that for a fixed value of $b = b^*$, the slope of the line segment $\Phi_n(b^*)$ is

$$-\frac{\sum_{i=1}^n (x_i - \bar{x})[x_i \delta_i + \sum_{k=1}^n w_{ik}(b^*) x_k (1 - \delta_i)]}{\sum_{i=1}^n (x_i - \bar{x})^2}.$$

10.3. Assume that the Kaplan-Meier product-limit estimator $\widehat{S}(u)$ satisfies the following self-consistency criterion

$$\widehat{S}(u) = \frac{1}{n} \left[N(u) + \sum_{i:t_i \leq u} (1 - \delta_i) \frac{\widehat{S}(u)}{\widehat{S}(t_i)} \right],$$

where $N(u)$ denotes the number of censored and uncensored observations exceeding u. In the Buckley-James method, suppose that censored residuals c_i are re-weighted in the scatterplot by weights w_{ik}, where, for each $i = 1, 2, \ldots, n, \; j = 1, 2, \ldots, n,$

$$w_{ik} = \begin{cases} \frac{v_k}{\widehat{S}(c_i)} \delta_k (1 - \delta_i) & \text{if } e_k > c_i \\ 0 & \text{otherwise}, \end{cases}$$

and $e_k = \min\{z_k, c_k\}$ represent observed residuals, both censored and uncensored ranked from smallest to largest; $v_k \delta_k$ is the size of the discontinuity in $\widehat{S}$ at e_k.

(a) By considering the definition of $v_k \delta_k$ and the self-consistency criterion, show that

$$nv_k\delta_k = 1 + \sum_{i=1}^{n} w_{ik}.$$

(b) Demonstrate that the equation in (a) holds for the $n = 4$ ranked residuals

e_k	−1	0.34	0.45	1.1
δ_k	0	0	1	1

10.4. For the simple linear model

$$Y_i = \beta_0 + \beta x_i + \epsilon_i, \quad i = 1, 2, 3, \ldots, n,$$

where the ϵ_i have mean 0 and variance σ^2, the t_i are fixed censor constants with

$$\delta_i = \begin{cases} 1 & \text{if } y_i < t_i \\ 0 & \text{if } y_i \geq t_i, \end{cases}$$

show that

$$Var[Y_i\delta_i + E(Y_i|Y_i > t_i)(1 - \delta_i)] = \sigma^2 - Var(\epsilon_i|\epsilon_i > c_i)S(c_i),$$

where $c_i = t_i - \beta x_i$ and the $\epsilon_i = Y_i - \beta x_i$ are independent and identically distributed with mean β_0, variance σ^2 and common survival function S.

10.5. Show that, for the censored simple linear regression method of Koul, Susarla and Van Ryzin,

$$E[\frac{\delta_i Z_i}{G(Z_i)}] = \beta_0 + \beta_1 x_i$$

when responses Y_i are randomly right-censored by random censor times T_i with survival function G and $Z_i = \min\{Y_i, T_i\}$.

10.6. Construct the Buckley-James matrix of weights $\mathcal{W}$ for a simple linear regression with five data points, two of which are censored. How many different weights matrices are there in this case?

Show that in each case

$$\sum_{k=1}^{5} w_{ik} = 1.$$

What does this equation say concerning the matrix $\mathcal{W}$?

10.7. Show that each of the possible 5×5 renovation weights matrices $\mathcal{W}$ satisfies: $\mathcal{W}^2 = \mathcal{W}$; $(I - \mathcal{W})^2 = I - \mathcal{W}$.

10.8. Assume that if Y is a survival random variable with survival function S and Y is right-censored by the fixed constant t, then

$$E(Y|Y > t) = t + \int_t^\infty \frac{S(y)}{S(t)} dy.$$

EXERCISES

(a) Use the above result to deduce that

$$E(Y) = \int_0^\infty S(y)dy.$$

(b) The following data gives the observed survival times (in weeks) of leukemia patients in a clinical trial where the patients enter the trial at time of diagnosis: 5, 5+, 6, 9, 9+, 10, 15.

  (i.) Construct the Kaplan-Meier product-limit estimator for this data using Efron's redistribute-to-the-right algorithm.
  (ii.) Estimate the probability of survival beyond 8 weeks after diagnosis.
  (iii.) Estimate the mean life expectancy from time of diagnosis.
  (iv.) Estimate the mean life expectancy of leukemia patients 8 weeks after diagnosis.

(c) The estimate of slope in a simple linear regression with data $(x_i, Y_i^*), i = 1, 2, 3, \ldots, n$ is defined by

$$\widehat{\beta} = \frac{\sum_{i=1}^n (x_i - \bar{x})Y_i^*}{\sum_{i=1}^n (x_i - \bar{x})^2}.$$

If $Y_i^* = Y_i \delta_i + E(Y_i|Y_i > t_i)(1 - \delta_i)$ when, for each i, independent and identically distributed survival times Y_i, with means $E(Y_i) = \beta_0 + \beta_1 x_i$, are respectively right-censored by a fixed censoring constants t_i, show that

$$E(\widehat{\beta}) = \beta_1.$$

What is the implication of this result for regression with censored data?

10.9. In fitting a Buckley-James simple linear regression to a scatterplot of eight points, the final ranking of the observed residuals from smallest to largest is given by

x o x o x o x x

where x denotes an uncensored residual, o denotes a censored residual and there are no ties. For this ordering we list the residuals as $r_1, r_2, \ldots, r_8$ and these comprise the residual vector **r**.

Let $\widehat{S}$ denote the product-limit estimator of the survival function of random variable R based on the observed values $r_1, r_2, \ldots, r_8$ of R.

(a) Use Efron's redistribute-to-the-right algorithm to determine the probability masses $v_1, v_2, \ldots, v_8$ assigned by the product-limit estimator to the eight ordered residuals $r_1, r_2, \ldots, r_8$.
(b) Determine $\widehat{S}(r_2)$, $\widehat{S}(r_4)$ and $\widehat{S}(r_6)$.
(c) Calculate the entries of the weight matrix $\mathcal{W}$.
(d) Use the weight matrix to estimate $E(R|R > r_2)$ as a linear combination of $r_1, r_2, r_3, \ldots, r_8$.
(e) Express $\mathcal{W}\mathbf{r}$ more simply in terms of **r**.

10.10. Show that the renovated residuals given by $\epsilon^* = Y^* - \mathbf{X}\widehat{\beta_n}$ produce the same renovated scatterplot (X, Y^*) as location-shifted residuals $\epsilon^* + c\mathbf{1}$, where c is a scalar.

10.11. Show that $\widehat{\mathbf{Y}^*}^T \epsilon^* = 0$. What is the implication of this result in terms of the geometry of the residuals? What should we seek in a plot of renovated residuals versus fitted values?

10.12. The Stanford Heart Transplant data listed in Example 5.2 gives observations on 'Days', being survival time (in days) after transplant; also listed is the right-censoring information for each patient under the heading 'Cens'; age of patient 'Age'; mismatch score, 'T5', a measure of how close alignment of the donor and recipient tissue.

(a) Fit a linear model to these data, taking $\log_e$(Days) as the response variable and include 'Age' and 'T5' as covariates contributing to the model.

(b) Compare the model fitted in (a) with the nested models that fit 'Age' and 'T5', respectively. Does it appear that 'Age' is the only significant covariate? Explain.

10.13. When fitting the simple linear model (10.2) using the Buckley-James method for right-censored responses, to $n = 4$ data points, suppose that for a given value of the slope b, the residuals (and their corresponding censor indicator and covariate) are ranked $e_1(b) < e_2(b) < e_3(b) < e_4(b)$ and that this residual ranking does not change for any value slope value in an interval I containing b. Assume that in this rank order, the corresponding censor indicators are $\boldsymbol{\delta} = (0, 1, 0, 1)^T$ and the corresponding single covariate $\mathbf{x}$ has values $\mathbf{x} = (1, 0, -k - 1, k)^T$.

(a) Determine the weights matrix for these data.

(b) Show that if $\Phi_n(b) = A_1 - bA_2$ for all b in the interval I, then $A_2 > 0$. Hence show that Φ_n is not necessarily monotone decreasing in b.

10.14. Show that the weight matrix in the Buckley-James method satisfies the following property:
$$\frac{1}{n}\mathbf{1}^T \mathcal{W} = \mathbf{v}^T,$$
where $\mathbf{v}$ is the vector of probability masses assigned by the product-limit estimator to the uncensored residuals.

10.15. The following data from Stablein and Koutrouvelis (1985) give results from a clinical trial (Gastrointestinal Tumor Study Group, 1982) comparing two treatments: chemotherapy versus chemotherapy combined with radiotherapy for gastric cancer patients. In the trial, 45 patients were randomised to each of the two treatments and monitored for 8 years. The data are measured in days and * denotes a right-censored observation. These data were discussed in Klein and Moeschberger (1997) in the context of censored data analogues of the Kolmogorov-Smirnov statistic.

			Chemotherapy only					
1	63	105	129	182	216	250	262	301
301	338	342	354	356	358	380	383	383
394	408	460	489	499	523	524	535	562
569	675	676	748	778	786	797	955	968
1000	1245	1271	1420	1551	1694	2363	2754*	2950*

			Chemotherapy and radiotherapy					
17	42	44	48	60	72	74	95	103
108	122	144	167	170	183	185	193	195
197	208	234	235	254	307	315	401	445
464	484	528	542	547	577	580	795	855
1366	1577	2060	2412*	2486*	2796*	2802*	2934*	2988*

(*Source*: Stablein, D.M. and Koutrouvelis, I.A. (1985). A two-sample test sensitive to crossing hazards in uncensored and singly censored data. *Biometrics*, 41, 643–52.)

Fit a linear model to these right-censored data using the Buckley-James method. Use group membership as a single covariate and $\log_e(\text{days})$ as the response. Is the zero-crossing exact? Examine a diagnostic plot of the renovated residuals $\epsilon_i = Y_i^* - \widehat{\beta_1} x_i$ versus the 'fitted values' $\widehat{\beta_1} x_i$ to seek violations to the assumption of constant residual variance in the model. Comment.

10.16. For simple linear regression with censored data, we assume that the Buckley-James estimators $\widehat{\beta_0}$ for intercept and $\widehat{\beta_1}$ for slope satisfy the modified normal equation

$$\sum_{i=1}^{n}(Y_i^* - \widehat{\beta_0} - \widehat{\beta_1} x_i) = 0.$$

Use this equation, and the residuals $e_i = z_i - \widehat{\beta_1} x_i$, to show that once $\widehat{\beta_1}$ is known, then $\widehat{\beta_0}$ is the mean of:

(a) the renovated residuals $\epsilon_i = Y_i^* - \widehat{\beta_1} x_i$;

(b) the censored and uncensored residuals

$$e_i \delta_i + \sum_{k=1}^{n} w_{ik} e_k (1 - \delta_i);$$

(c) $\widehat{F}(e_k)$.

10.17. Show that the rows of the weights matrix defined by (10.22), used to construct weights in the Buckley-James method for interval-censored data, still sum to 1.

Show that under the special case of right-censoring, the weights matrix for interval-censoring defined by (10.22) simplifies to the weights matrix of (10.10).

CHAPTER 11

Buckley-James Diagnostics and Applications

11.1 Censored regression diagnostics

According to Definition 10.4, once a Buckley-James estimator $\mathbf{b} = \widehat{\beta_n}$ has been found, the response data may be 'renovated' to construct a scatterplot of new responses. In practice, this implementation happens after the estimator has been found. It allows the renovation process to occur even when $\Phi_n(\mathbf{b})$ undergoes a zero-crossing at $\mathbf{b} = \widehat{\beta_n}$ rather than an exact solution. The process for implementation is outlined in Smith and Peiris (1999) as follows. As usual, let

$$\mathcal{W} = \mathcal{W}(\widehat{\beta_n})$$

denote the weight matrix based on the Buckley-James solution. Using this value of $\mathcal{W}$, we create the slope estimate

$$\widehat{\beta} = (\mathbf{X}^T \mathcal{W} \mathbf{X})^{-1} \mathbf{X}^T \mathcal{W} \mathbf{Z},$$

which we conveniently term the **renovation slope estimate** of β, where the general use of the word 'slope' is fashioned from two-dimensional scatterplots. This allows us to define the scatterplot renovation equation for the renovated responses $\mathbf{Y}^* = (Y_1^*, Y_2^*, \ldots, Y_n^*)^T$ by

$$\mathbf{Y}^* = \mathbf{X}\widehat{\beta} + \mathcal{W}(\mathbf{Z} - \mathbf{X}\widehat{\beta}). \tag{11.1}$$

Since $\mathcal{W}\mathbf{Y}^* = \mathcal{W}\mathbf{Z}$, it follows that the renovated scatterplot satisfies

$$\widehat{\beta} = (\mathbf{X}^T \mathcal{W} \mathbf{X})^{-1} \mathbf{X}^T \mathcal{W} \mathbf{Y}^* \tag{11.2}$$

$$\mathbf{Y}^* = \mathbf{X}\widehat{\beta} + \mathcal{W}(\mathbf{Y}^* - \mathbf{X}\widehat{\beta}). \tag{11.3}$$

We assume this structure for $\mathcal{W}, \widehat{\beta}$ and $\mathbf{Y}^*$ in what follows. As in the previous chapter, we regard a **renovated scatterplot** as a diagrammatic depiction of the pair $(\mathbf{X}, \mathbf{Y}^*)$. Note that equation (10.16) and the result of Theorem 10.6 (cf. Smith and Zhang, 1995; Currie, 1996) show that

$$\widehat{\beta_n} = \widehat{\beta} \Leftrightarrow \Phi(\widehat{\beta_n}) = 0,$$

so that the Buckley-James estimate is equal to the renovated slope estimate of β if and only if the zero-crossing is exact.

We now consider some standard diagnostic techniques (see, for example,

Belsley, Kuh and Welsch, 1980) applied to the renovated scatterplot and the renovation slope estimator $\hat{\beta} = (\mathbf{X}^T \mathcal{W} \mathbf{X})^{-1} \mathbf{X}^T \mathcal{W} \mathbf{Y}^*$ based on the Buckley-James estimator. This formula is useful since it shows how the covariates associated with uncensored points, through $\mathcal{W}$, are used to affect the positions of the censored points in the renovation process. Thus, if an uncensored observation is unusual in some sense (high leverage, influential, large residual, ...), it may have an effect on the fitting process in the renovated scatterplot, both through its own location, and by its role in the positioning of censored neighbours.

Definition 11.1 *The* **fitted values** *in the renovated scatterplot are given by* $\widehat{\mathbf{Y}^*} = \mathbf{X}\hat{\beta}$. □

According to Definition 11.1, the fitted values may be written in the form

$$\widehat{\mathbf{Y}^*} = \mathbf{X}\widehat{\beta_n} = \mathbf{X}(\mathbf{X}^T \mathcal{W} \mathbf{X})^{-1} \mathbf{X}^T \mathcal{W} \mathbf{Y}^* = \mathbf{H}^* \mathbf{Y}^*,$$

where $\mathbf{H}^* = \mathbf{X}(\mathbf{X}^T \mathcal{W} \mathbf{X})^{-1} \mathbf{X}^T \mathcal{W}$ is a special matrix containing important diagnostic information. It is, in fact, the censored regression equivalent of the **hat matrix** $\mathbf{H} = \mathbf{X}(\mathbf{X}^T \mathbf{X})^{-1} \mathbf{X}^T$ from classical regression.

Definition 11.2 *The* **renovated hat matrix**, $\mathbf{H}^*$, *is given in terms of the weights matrix* $\mathcal{W}$ *by* $\mathbf{H}^* = \mathbf{X}(\mathbf{X}^T \mathcal{W} \mathbf{X})^{-1} \mathbf{X}^T \mathcal{W}$. □

$\mathbf{H}^*$ has the following easily-established properties of a 'projection' matrix, apart from the symmetry (which is present in weighted least squares but not in renovated least squares).

Theorem 11.1 *The renovated hat matrix* $\mathbf{H}^*$ *satisfies:*

1. $\mathbf{H}^*$ *and* $\mathbf{I} - \mathbf{H}^*$ *are idempotent;*
2. $trace(\mathbf{H}^*) = p$ *and* $trace(\mathbf{I} - \mathbf{H}^*) = n - p$, *for* $\mathbf{X}$ *of full rank* p;
3. $\mathbf{H}^* \boldsymbol{\epsilon}^* = 0$, *where* $\boldsymbol{\epsilon}^* = \mathbf{Y}^* - \mathbf{X}\hat{\beta}$, *the vector of renovated residuals.*

□

The **renovated leverages** or **renovated HI values** comprise the diagonal entries of $\mathbf{H}^*$. They may be expressed in terms of $\mathbf{w_i}$, the ith column of $\mathcal{W}$, and $\mathbf{x_i}^T$, the ith row of $\mathbf{X}$ by

$$h_{ii}^* = \mathbf{x_i}^T (\mathbf{X}^T \mathcal{W} \mathbf{X})^{-1} \mathbf{X}^T \mathbf{w_i}. \tag{11.4}$$

Theorem 11.1 suggests that the average size of the leverage values is still

$$\frac{trace(\mathbf{H}^*)}{n} = \frac{p}{n},$$

so that points with leverages exceeding $\frac{2p}{n}$ could be flagged as unusually large. Since

$$h_{ij}^* = \mathbf{x_i}^T(\mathbf{X}^T\mathcal{W}\mathbf{X})^{-1}\mathbf{X}^T\mathbf{w_j}$$

measures the influence of the jth point on the fit at the ith point when the zero-crossing is exact, then if $\delta_j = 0$ and the jth point is censored, $\mathbf{w_j} = \mathbf{0}$ and $h_{ij}^* = 0$ and $h_{jj}^* = 0$ as expected.

Since

$$trace(\mathbf{H}^*) = trace(\mathbf{H}) = p,$$

where $\mathbf{H} = \mathbf{X}(\mathbf{X}^T\mathbf{X})^{-1}\mathbf{X}^T$ is the usual hat matrix, it follows that because censored points have no renovated leverage, ($h_{jj}^* = 0$), uncensored points compensate with increased renovated leverage so as to maintain the constancy of the trace of the renovated hat matrix. This increases the need and importance of checking renovated leverage values for points in the renovated scatterplot. This is demonstrated in Example 11.1.

If the ith (uncensored) observation is deemed to exert unusually large influence on the least squares fit, some measure of the effect on the slope estimator may be obtained through **deletion formulas**. (See, for example, Myers, 1990.) We are less concerned about the deletion of a censored observation, since its position in the renovated scatterplot depends on its uncensored neighbours, and it is these neighbours which may be exerting high influence. The following results are due to Smith and Zhang (1995).

Suppose that $\delta_i = 1$. Let $\mathbf{X}_{(i)}$ denote the $\mathbf{X}$ matrix with the ith row removed and $\mathcal{W}_{(i)}$ the matrix derived from $\mathcal{W}$ by deleting the ith row $(\mathbf{0}^T, 1, \mathbf{0}^T)$ and ith column $\mathbf{w_i} = (\mathbf{a}^T, 1, \mathbf{0}^T)^T$. By writing

$$\mathcal{W}_{(i)} = \begin{pmatrix} \mathbf{A}_{11} & \mathbf{A}_{12} \\ \mathbf{A}_{21} & \mathbf{A}_{22} \end{pmatrix}$$

and performing block multiplication on the partitioned matrix structure

$$\mathbf{X}^T\mathcal{W}\mathbf{X} = \left(\mathbf{X_1}^T \vdots \mathbf{x_i} \vdots \mathbf{X_2}^T\right) \begin{pmatrix} \mathbf{A}_{11} & \vdots & \mathbf{a} & \vdots & \mathbf{A}_{12} \\ \cdots & & \cdots & & \cdots \\ \mathbf{0}^T & \vdots & 1 & \vdots & \mathbf{0}^T \\ \cdots & & \cdots & & \cdots \\ \mathbf{A}_{21} & \vdots & 0 & \vdots & \mathbf{A}_{22} \end{pmatrix} \begin{pmatrix} \mathbf{X_1} \\ \cdots \\ \mathbf{x_i}^T \\ \cdots \\ \mathbf{X_2} \end{pmatrix},$$

it follows that $\mathbf{X}_{(i)}^T\mathcal{W}_{(i)}\mathbf{X}_{(i)} = \mathbf{X}^T\mathcal{W}\mathbf{X} - \mathbf{X}^T\mathbf{w_i}\mathbf{x_i}^T$. Thus, according to the inversion formula of Rao (1973),

$$(\mathbf{X}_{(i)}^T\mathcal{W}_{(i)}\mathbf{X}_{(i)})^{-1} = (\mathbf{X}^T\mathcal{W}\mathbf{X})^{-1} + \frac{(\mathbf{X}^T\mathcal{W}\mathbf{X})^{-1}\mathbf{X}^T\mathbf{w_i}\mathbf{x_i}^T(\mathbf{X}^T\mathcal{W}\mathbf{X})^{-1}}{1 - \mathbf{x_i}^T(\mathbf{X}^T\mathcal{W}\mathbf{X})^{-1}\mathbf{X}^T\mathbf{w_i}}.$$

Finally, if $\widehat{\beta}_{(i)} = (\mathbf{X}_{(i)}^T\mathcal{W}_{(i)}\mathbf{X}_{(i)})^{-1}\mathbf{X}_{(i)}^T\mathbf{Y}_{(i)}^*$ is the deleted slope estimate when the ith point $(\mathbf{x_i}, y_i^*)$ is deleted from the renovated scatterplot, then these matrix inversion results show that

$$\widehat{\beta} - \widehat{\beta}_{(i)} = \frac{(\mathbf{X}^T\mathcal{W}\mathbf{X})^{-1}\mathbf{X}^T\mathbf{w_i}\epsilon_i^*}{1 - h_{ii}^*}, \qquad (11.5)$$

where $\epsilon_i^* = y_i^* - \mathbf{x_i}^T\widehat{\beta}$.

Similarly, when $\delta_i = 0$ and the ith observation is censored, then (11.5) still holds with the roles of $\mathbf{X}^T\mathbf{w_i}$ and $\mathbf{x_i}$ interchanged, since the non-zero entries in the weights matrix now occur in the ith row rather than the ith column. In this case, the matrix inversion formulas are based on

$$\mathbf{X}_{(i)}{}^T\mathcal{W}_{(i)}\mathbf{X}_{(i)} = \mathbf{X}^T\mathcal{W}\mathbf{X} - \mathbf{x_i}\mathbf{w_i}^T\mathbf{X},$$

where $\mathbf{w_i}^T$ is the ith row of $\mathcal{W}$.

Notice that $\widehat{\beta}_{(i)}$ is not the slope if the Buckley-James renovation process is applied to the original data set less the point $(\mathbf{x_i}, y_i^*)$ since the entries of $\mathcal{W}$ are functions of the sample size. That is, $\widehat{\beta}_{(i)} \ne \widehat{\beta}$ based on a sample of size $n - 1$. Nevertheless, the deletion formula (11.5) is a useful first indicator of the effect of removal of, for example, the smallest residual (negative and of large magnitude), for then the weights assigned with a sample of size n and the weights assigned in the deleted sample differ only by a scale factor of $\frac{n-1}{n}$.

To forecast the effect of the **change in fit**, $dfit$, defined by

$$dfit = \mathbf{x_i}^T\widehat{\beta} - \mathbf{x_i}^T\widehat{\beta}_{(i)},$$

produced by deletion at the ith uncensored point, it follows from (11.5) that

$$dfit = \frac{h_{ii}^*\epsilon_i^*}{1 - h_{ii}^*}.$$

This is reduces to the usual least squares formula for 'change in fit' in the absence of censoring.

■ Example 11.1
Diagnostic applications to the Stanford Heart Transplant data

In the renovated scatterplot, the positions of censored points depend in a complex way on the positions of the uncensored points. This restricts the usefulness of the plot for inference, but does not restrict the usefulness of the plot as a visual diagnostic tool. We indicate this usefulness by reference to the Stanford Heart Transplant data ($n = 69$) from Example 5.2. The 'age' and 'logarithm (base 10) days' survival of the original 69 patients observed in this program between October 1967 and April 1974 are plotted in Figure 10.2, where 24 patients were censored at study termination. The vertical lines trace the paths of the censored patients in the renovation process. The diagram successfully gives an impression of the nature of a renovated scatterplot. The Buckley-James method gives unique solutions for the slope and intercept estimators, coincident with the fitted least squares line

$$\log \text{days} = 3.596 - 0.0278 \text{ age},$$

which shows significant slope (Buckley and James, 1979).

From Smith and Zhang (1995), a comparison of leverages and renovated leverages is given below for patients of various ages.

Age (x_i)	Censoring (δ_i)	Renovated leverage (h_{ii}^*)	Leverage (h_{ii})
19.7	1	0.347	0.145
23.7	0	0.000	0.109
52.0	1	0.022	0.021
64.5	1	0.116	0.078

The point $(19.7, \log_{10} 228)$, on the extreme left-side of the plot, representing the youngest uncensored patient, exhibits high renovated leverage in the plot; the renovated leverage exceeding the leverage and causing a change in fit of $dfit = -0.366$, so that the slope is flatter when this patient is included and more negative when this patient is excluded.

□

11.2 Plots of renovated data

In this section we apply the Buckley-James method for censored regression to the p-sample problem where the samples are subject to right-censoring. We follow the description given in Smith (1995) for constructing renovated scatterplots in the two-sample problem. The right-censored samples are reconstructed so as to remove the effect of censoring and graphical procedures based on quantiles (such as boxplots) may then be used as a standard data-analytic tool to describe the variable being measured.

For example, the $p = 2$ sample problem uses a covariate

$$x_i = \begin{cases} 1 \text{ if } z_i \text{ is from Sample 1} \\ 0 \text{ if } z_i \text{ is from Sample 2} \end{cases}$$

with the linear model as before. Once a Buckley-James solution $\widehat{\beta_n}$ has been found in the least squares two-sample problem, then the data may be 'renovated': by a **renovated dotplot** we mean a plot of $(\mathbf{X}, \mathbf{Y}^*, \mathbf{\Delta})$, where

$$\mathbf{Y}^* = \mathbf{Y}^*(\widehat{\beta}) = \mathbf{X}\widehat{\beta} + \mathcal{W}(\mathbf{Z} - \mathbf{X}\widehat{\beta}),$$

$\mathcal{W} = \mathcal{W}(\widehat{\beta_n})$, $\widehat{\beta} = (\mathbf{X}^T \mathcal{W} \mathbf{X})^{-1} \mathbf{X}^T \mathcal{W} \mathbf{Y}^*$, and $\mathbf{\Delta}$ contains the plot symbols for uncensored points and renovated points. Standard boxplot comparisons may then take place on the renovated data $(\mathbf{X}, \mathbf{Y}^*, \mathbf{\Delta})$. Figure 11.1 shows a renovated scatterplot for leukemia data.

The renovated data may be usefully employed in QQ-plots to detect an underlying distribution for the response. Importantly, after renovation, for each of the p samples we may produce plots of the empirical survivor function $S_m(u)$, defined as the fraction of the m within-sample data exceeding u. When the linear model is appropriate, the consistency of the Buckley-James

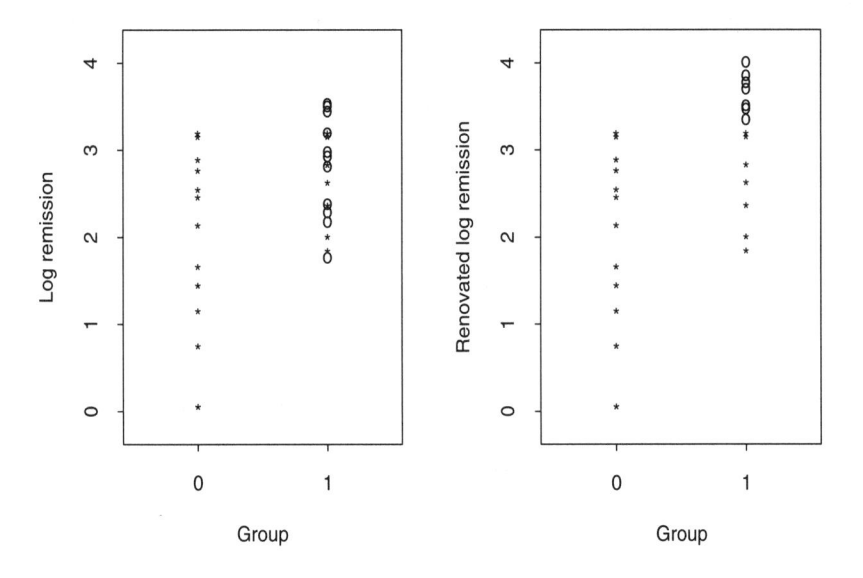

Figure 11.1 *Dotplots of original and renovated data for leukemia log remission times in Example 11.2.*

estimators implies that the renovated points will provide a guide to the shape of the survival function for each group.

However, it must be remembered that a necessary condition for the consistency of the slope parameter (James and Smith, 1984) is that the expected number of censored and uncensored residuals be large over the support of the residual distribution. We reinforce this mathematically: as in Chapter 10, if we define $\mathcal{N}_b(u)$ to be the expected number of censored and uncensored residuals $e_i(b)$, based on a given slope b, exceeding u, then as in (10.17), we require that

$$\mathcal{N}_b(u) \to \infty \text{ as } n \to \infty \text{ for all } u \in (-\infty, U_b],$$

where U_b lies at the top of the support of the residual distribution. This leads us to offer a word of caution. If we construct renovated scatterplots based on small overall sample sizes and unfavourable censoring patterns at the top of the residual distribution, even consistent estimates may produce unhelpful plots.

■ Example 11.2
Renovating and plotting data for the general two-sample problem: comparative dotplots for leukemia remissions

Lawless (1982), Gehan (1965) and others have discussed data from a clinical trial examining steroid induced remission times (weeks) for leukemia patients. One group of 21 patients was given 6-mercaptopurine (6-MP); a second group of 21 patients was given a placebo. Since the trial lasted 1 year and patients

PLOTS OF RENOVATED DATA 221

were admitted to the trial during the year, right-censoring occurred at the cut-off date when some patients were still in remission. Observations $\log_e Z$ on log remission time $\log_e Y$ are given in the tables which follow. The table on the left-side shows the unrenovated data; in this table, '+' denotes right-censoring in the 6-MP group, and therefore represents remission which was still in effect at the closure of the trial. The table on the right-side shows the renovated data. It is particularly noticeable that uncensored observations are not moved in the renovation process.

6-MP unrenovated		Placebo unrenovated		6-MP renovated		Placebo renovated	
1.79	2.83+	0.00	2.08	1.79	3.72	0.00	2.08
1.79	2.94+	0.00	2.08	1.79	3.79	0.00	2.08
1.79	3.00+	0.69	2.40	1.79	3.79	0.69	2.40
1.79+	3.09	0.69	2.40	3.37	3.09	0.69	2.40
1.95	3.14	1.10	2.48	1.95	3.14	1.10	2.48
2.20+	3.22+	1.39	2.48	3.50	3.87	1.39	2.48
2.30	3.47+	1.39	2.71	2.30	4.02	1.39	2.71
2.30+	3.47+	1.61	2.83	3.53	4.02	1.61	2.83
2.40+	3.52+	1.61	3.09	3.53	4.02	1.61	3.09
2.56	3.56+	2.08	3.14	2.56	4.02	2.08	3.14
2.77		2.08		2.77		2.08	

The Buckley-James method provides an exact solution to the model parameters in this two-sample problem on the logarithm scale with covariate $x = 1$ for 6-MP; $x = 0$ for Placebo. The logarithmic transformation has the effect of stabilising the variance in the two groups being compared.

Dotplots of the original data and the renovated data are given in Figure 11.1, where multiple points have been overwritten by a single plot symbol (an asterisk for uncensored and a small open circle for censored). Notice that it is the smallest censor times which receive the greatest renovation; the censored observation at 1.79 log weeks is renovated to over 3.37 log weeks. In general, censored points with the most negative residuals are renovated toward the mean. The renovated dotplot is a view of the patients' survival without censoring. Such renovation produces a change of rank order in the data set, with wider spread apparent in the 6-MP data in Figure 11.1.

□

Having established the new ranks in the renovation process, boxplot comparisons of the two groups may proceed. The renovated data needs to be re-ranked as the censored points will all be larger. That is, the renovation process changes the order in the data. Boxplots are particularly visually appropriate for renovated data because such plots are based on ranks. In the renovated data, the censored responses are moved, in particular, to appropriate rank positions.

Classical boxplot techniques have been used for graphing censored data in a distribution-free way. The boxplot method of Gentleman and Crowley (1991, 1992) is based on inverting the product-limit estimator to locate the appropriate four quantiles in the data that are needed for boxplot display.

■ Example 11.3
Comparative boxplots for leukemia remissions — Example 11.2 continued

Figure 11.2, from Smith (1995), particularly shows boxplots for the 6-MP data where the right-censoring occurs. The Placebo data is uncensored and remains unchanged in the renovation process, so that standard boxplot techniques apply, with the resulting boxplot labelled 'Placebo' in Figure 11.2.

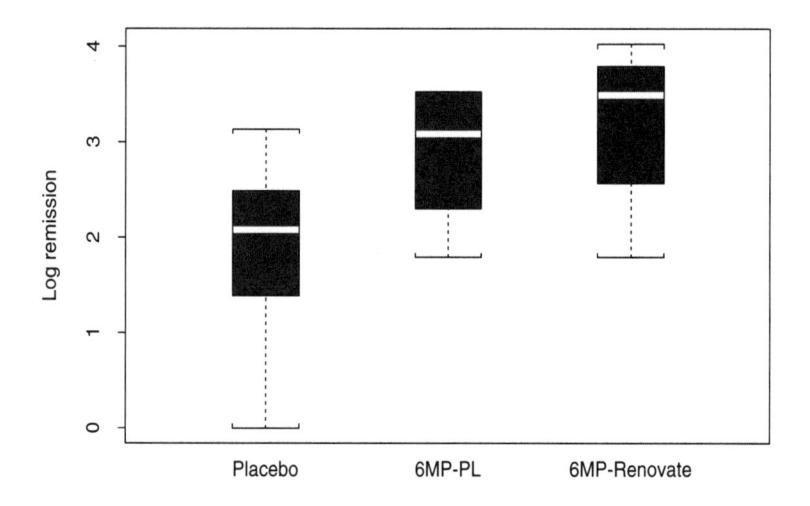

Figure 11.2 *Boxplots of renovated data for leukemia log remission times (Example 11.2) for 'Placebo' and '6MP-Renovate'; for comparison, '6MP-PL' is a Gentleman and Crowley (1992) boxplot of the 6-MP data based on the PL-estimator.*

In Figure 11.2, the inverse product-limit method of Gentleman and Crowley (1992), labelled '6MP-PL', is compared directly with the renovated boxplot labelled '6MP-Renovate'. The difference between the two plots is partially caused by the large proportion of censored data at the top of the 6-MP distribution leaving the product-limit estimator apparently 'hanging' before its conventional assignment to zero beyond the largest observation. This leaves the boxplot without an upper whisker. On the other hand, in the 6MP-Renovate plot, the data have been renovated to estimated lifetimes reflected by an upper whisker in the boxplot. Clearly, these plotting techniques are more effective for larger sample sizes. It must always be remembered that a condition necessary for the consistency of the Buckley-James estimators is that, asymptotically, the expected number of censored residuals and uncensored residuals be large over the support of the residual distribution. □

■ Example 11.4
Comparative plots of the PL-estimator and empirical survivor function for renovated leukemia remissions

Once data have been renovated through the Buckley-James procedure, then simple methods such as plotting the empirical survivor function of the renovated data may produce insight into the shape of the underlying survival function.

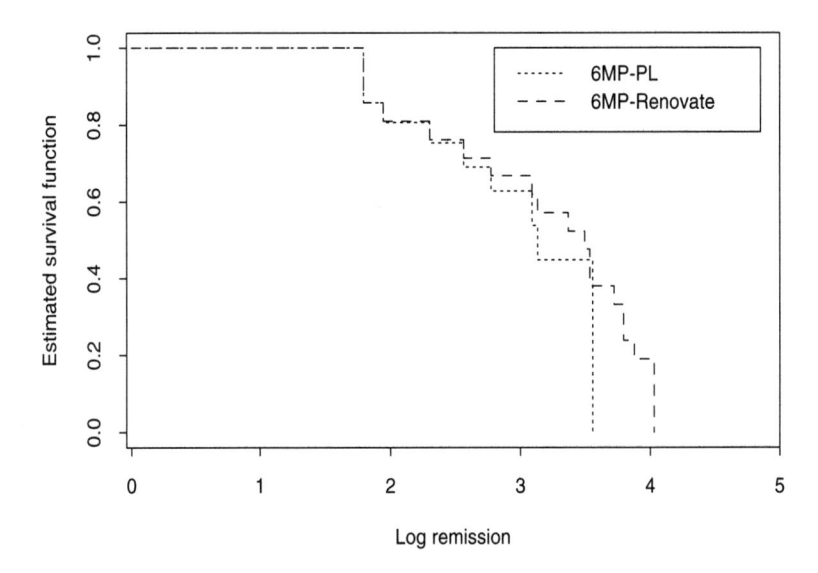

Figure 11.3 *Comparisons for the 6-MP group: empirical survivor function for renovated log lifetime (dashed lines); product-limit estimator for log lifetime (dotted lines).*

This effect is demonstrated in Figure 11.3, where only the 6-MP data are plotted on the log scale. The plot of the standard product-limit estimator $\widehat{S}^*$ is represented by dotted lines and is an estimator of S^*, the survival function for log remission. On the same log scale, the empirical survivor function S_{21}^* is represented by dashed lines. This is the familiar estimator defined by (4.1) having discrete jumps of $\frac{1}{n}$ for each (renovated) data point plotted.

As a consequence of the final residual rankings of both groups combined, the renovation process moves some censored times beyond the largest pre-renovation observation in the 6-MP group, thereby reducing the 'hanging effect' of the estimated survival function at the top of the distribution. For the empirical survivor function, the points of discontinuity occur at every distinct renovated data point. In comparison, because of the redistribute-to-the-right algorithm (Section 6.4), the jump sizes at the points of discontinuity of $\widehat{S}^*$ increase toward the top of the distribution.

Notice that generally S_m^* depends on the censoring pattern in both the samples being compared — sample membership is needed as a covariate in the Buckley-James regression — whereas $\widehat{S}^*$ is determined from a single sample. However, when the linear model is appropriate, the consistency of the Buckley-James estimators implies that, provided that the expected number of censored observations and uncensored observations is large over the support of the survival distribution, both $\widehat{S}^*$ and S_m^* are uniformly consistent estimators of the same survival function. For moderate sample sizes, when the linear model is appropriate, the graph of the empirical survivior function of the renovated data provides an alternative to the graph of the product-limit estimator on the observed data. □

11.3 Functions for renovating censored responses

In this section we provide S-PLUS functions for estimating the magnitude and rank positions of response data that have been right-censored in the presence of one or more explanatory variables. This process of sample adjustment, or renovation, allows samples to be compared graphically, using diagrams such as boxplots which are based on ranks as in Section 11.2. The renovation process is based on Buckley-James censored regression estimators. As part of their output, the functions provide iterative solutions to the Buckley-James estimating (normal) equations in an interactive manner by minimising the euclidean norm at adjacent iterates. We list the S-PLUS functions as:

S-PLUS functions

`ren.fit()`	for estimates in the Buckley-James method and for renovating the right-censored data
`weight()`	for determining the weight matrix at each iteration
`sur1()`	for appropriate management of tied data
`sur2()`	for counting censored points in tied data

The functions are for determining the Buckley-James estimators for β and $E(\epsilon)$ in the linear model

$$\mathbf{Y} = \mathbf{X}\beta + \epsilon \qquad (11.6)$$

introduced in (10.12). Notice that (11.6) may be applied to both univariate and multivariate regression data; in the univariate case $\mathbf{X}$ is a vector, while in the multivariate case $\mathbf{X}$ is a matrix. Both cases may be analysed using the same renovation function. The Buckley-James estimators result from a 'fit' of (11.6) to right-censored data. The renovated data $\mathbf{Y}^*$, which results from the

FUNCTIONS FOR RENOVATING CENSORED RESPONSES

final iteration of such a fit, is also provided as output. Plots of renovated data may then be simply constructed.

A main function called `ren.fit` carries out the processes of estimation and renovation by cycling through an iterative scheme indexed by s.

Input into `ren.fit`:

S-PLUS code	Definition	Symbol	Dimension
z	observed response	**Z**	vector
status	censor indicators	δ	vector
x	covariate(s)	**X**	vector/matrix
num	maximum number of iterations		integer

For $s = 0$, an initial estimate $\mathbf{b}_0$ of $\boldsymbol{\beta}$ is found from the least squares fit through the uncensored responses. For $s = 1, 2, 3, \ldots$, at iteration $s+1$, the current estimate $\mathbf{b}_s$ of $\boldsymbol{\beta}$ is updated via a least squares estimate $\mathbf{b}_{s+1}$ through new responses $\mathbf{Y}^*(\mathbf{b}_s) = \mathbf{X}\mathbf{b}_s + \mathcal{W}(\mathbf{b}_s)(\mathbf{Z} - \mathbf{X}\mathbf{b}_s)$. This formula necessitates, at each iteration, a call to the weight matrix function `weight` to determine $\mathcal{W}(\mathbf{b}_s)$ for the current value of $\mathbf{b}_s$. Since the weight matrix is carefully constructed from the product limit estimator applied to the ranked residuals, the function adopts conventions usual to the construction of this estimator: at each iteration, the largest observed residual is deemed to be uncensored for calculations at that iteration; tied uncensored residuals precede censored residuals with which they are tied. Procedures for counting and breaking ties are resolved by calls to functions `sur1` and `sur2`. Ties occur frequently in situations where a covariate represents group membership.

Output from `ren.fit`:

S-PLUS code	Definition	Symbol	Dimension
oldresponse	observed response	**Z**	vector
response	renovated response	$\mathbf{Y}^*(\widehat{\boldsymbol{\beta}_n})$	vector
censorindicator	censor indicators	δ	vector
covariate	covariate(s)	**X**	vector/matrix
bjslope		$(\widehat{\alpha_n}, \widehat{\boldsymbol{\beta}_n})$	vector/matrix
lsslope		$(\widehat{\alpha_{LS}}, \widehat{\boldsymbol{\beta}_{LS}})$	vector/matrix
trace		$\mathbf{b}_s, s = 1, 2, 3, \ldots$	vector/matrix
norm	euclidean norms	$\|\mathbf{b}_s - \mathbf{b}_{s-1}\|$, $s = 1, 2, \ldots$	vector

Featured in the output from `ren.fit` are: the covariate(s) used; the responses before and after renovation; the censor indicators rearranged in response order; Buckley-James estimate of $\alpha = \beta_0$ and $\boldsymbol{\beta} = (\beta_1, \beta_2, \ldots, \beta_{p-1})^T$, least squares estimates $(\widehat{\alpha_{LS}}, \widehat{\boldsymbol{\beta}_{LS}})$ through a least squares fit to the renovated scatterplot. These final slope estimates may be seen in the context of a summary of the iteration process: slope estimates $\mathbf{b}_s$ are given for each s and, as a measure of convergence, the euclidean norm of $\mathbf{b}_s - \mathbf{b}_{s-1}$ for $s = 1, 2, 3, \ldots$.

Specific features of `ren.fit` are: variable MULTI assesses whether a univariate or multivariate version of (11.6) is being fitted; the iteration number s is held by icnt; the iterations break when $||\mathbf{b}_s - \mathbf{b}_{s-1}|| < 0.0001$ or when s reaches num.

S-PLUS coding for the renovation function — ren.fit()

```
ren.fit<-function(z, status, x, num) {
# if the covariate is a matrix,
# set MULTI=1; else MULTI=0. if MULTI=1,
# then, set the dimension, 'dmn' to 2;
# else number of covariates plus 1.
# set all initial parameters for the loop input.
        MULTI <- 0
        if(is.matrix(x))
                MULTI <- 1
        if(MULTI == 0) {
                dmn <- 2
        }
        else {
                dmn <- length(x[1, ]) + 1
        }
        if(num == 1)
                enorm <- 0
        if(num >= 2)
                enorm <- array(0, num - 1)
        if(MULTI == 0) {
                timex <- array(0, length(x))
        }
        else {
                timex <- matrix(0, dmn - 1, length(x[, 1]))
        }
        timex <- x
        timez <- z
        slope <- lsfit(timex, timez)$coef[-1]
#initialise 'num x dmn' matrix 'slopebj'.
        slopebj <- matrix(0, num, dmn)
#the loop commences here.
        for(icnt in 1:num) {
```

FUNCTIONS FOR RENOVATING CENSORED RESPONSES

```
#compute the residual vector appropriately.
            if(MULTI == 0) {
                    r <- timez - (timex * slope)
            }
            else {
                    r <- timez - (timex %*% slope)
            }
            ord <- order(r)
#arrange the covariate vector or matrix
#in the order of residuals.
            if (MULTI==0){
            timex<-timex[ord]
            }
            else {
            timex<-timex[ord,1:length(x[1,])]
            }
            timez <- timez[ord]
            status <- status[ord]
            r <- r[ord]
#add the absolute minimum value of the residuals
#to avoid negative values.
            absminr <- abs(min(r))
            rr <- r + absminr
#go to weightmatrix computation.
            wm <- weight(rr, timex, timez, status)
#retrieve the output from weightmatrix routine
#and make necessary adjustments.
            r <- as.vector(wm$r)
            if(MULTI == 0) {
                    timex <- as.vector(wm$timex)
            }
            else {
                    timex <- as.matrix(wm$timex)
            }
            timez <- as.vector(wm$timez)
            status <- as.vector(wm$status)
            r <- r - absminr
            w <- as.matrix(wm$wm)
#compute new response vector.
            if(MULTI == 0) {
                    z <- (timex * slope) + w %*% r
            }
            else {
                    z <- (timex %*% slope) + w %*% r
            }
```

```
                        z <- as.vector(z)
#recompute the least squares fit.
            b <- lsfit(timex, z)$coef
            slopebj[icnt, ] <- b
            slope <- b[-1]
#if the loop is repeated, compute euclidean norm between
# consecutive slope estimates and save in 'enorm'.
            if(num == 1)
                    break
            if(num >= 2)
            enorm[icnt - 1] <- sqrt(sum((slopebj[icnt, -1] -
                            slopebj[icnt - 1, -1])^2))
#if if 'enorm' stays below 0.0001, stop iterations.
            if(icnt >= 2 && enorm[icnt - 1] < 0.0001)
                    break
     }
     if(MULTI == 0) {
            r <- z - (timex * slope)
     }
     else {
            r <- z - (timex %*% slope)
     }
     intercept <- mean(r)
     bj <- c(intercept, slope)
     bls <- lsfit(timex, z)$coef
#write output to a list.
     list(covariate = timex, oldresponse = timez,
            response = z, censorindicator = status,
            bjslope = bj, lsslope = bls, trace = slopebj,
            norm = enorm)
}
```

The following function **weight** is called by the main renovation function **ren.fit**. At iteration s, **weight** calculates $\mathcal{W}(\mathbf{b}_s)$ using the product-limit estimator applied to the residuals $E(\mathbf{b}_s) = \mathbf{Z} - \mathbf{Xb}_s$ which are made positive (so as to behave like survival data) by adding to all entries the absolute value of the smallest, and sorted from smallest to largest. This requires $E(\mathbf{b}_s)$, $\mathbf{Z}$, $\mathbf{X}$ and δ as input to the function.

Calls to **sur1** and **sur 2** are used to carefully count ties and check that the largest observed response is uncensored at each iteration. At each iteration, the weight matrix is calculated from the formula (10.10) using the 'outer product' of two vectors: the **outer product of vectors** $\mathbf{a} = (a_1, a_2, \ldots, a_n)^T$ and $\mathbf{b} = (b_1, b_2, \ldots, b_n)^T$ is the matrix with (i,j)th entry $a_i b_j$. That is,

$$\text{outer}(\mathbf{a}, \mathbf{b}) = [a_i b_j].$$

S-PLUS coding for the weight matrix — weight()

```
weight<-function(rr, timex, timez, delta) {
#FLAG checks for ties at the top of the residual distribution
#sur1 holds frequencies of distinct residuals and zero for
#remaining ties
        FLAG <- 0
        n <- length(rr)
        icount <- sur1(rr)
        icount <- icount[icount != 0]
#icount holds frequencies of distinct residuals with zero entries
#which were present in sur1 now removed.
#sur2 finds the frequencies of tied censored residuals
#and stores in inumb
        inumb <- sur2(icount, delta, n)
        if(icount[length(icount)] > 1) {
                FLAG <- 1
        }
#FLAG = 1 if ties at top of residual distribution
#such ties are broken by adding a small quantity to the largest
#giving radjd
#sur1, icount, sur2, inumb  recalculated using adjusted residuals
        radjd <- rr
        if(FLAG == 1){
                radjd[n] <- radjd[n] + radjd[n]/10
                icount <- sur1(radjd)
                icount <- icount[icount != 0]
                inumb <- sur2(icount, delta, n)
                }
#sort the input data in the order of the adjusted residuals
        o <- order(radjd, 2 - delta)
        presorted <- cbind(o, rr, delta, timez, radjd, timex)
        presorted <- as.matrix(presorted)
        sorted <- presorted[sort.list(presorted[, 1]),  ]
        rr <- sorted[, 5]
        delta <- sorted[, 3]
        status <- sorted[, 3]
        timez <- sorted[, 4]
        timex <- sorted[, -1:-5]
        delta <- c(delta[1:(n - 1)], 1)
#product-limit estimates calculated
        fit <- surv.fit(rr, delta)
        shat <- fit$surv
```

```
                v1 <- c(1, shat[1:(length(shat) - 1)])
                v <- v1 - shat
#icountv stores the probability mass for each data point
#including the ones for which delta takes a zero value.
            {
                k <- 0
                icounts <- array(0, n)
                icountv <- array(0, n)
                icount <- icount[icount != 0]
                num <- length(icount)
                if(num == n) {
                        icounts <- shat
                        icountv <- v
                }
                else {
                        for(i in 1:num) {
                                m <- icount[i]
                                for(j in 1:m) {
                                        k <- k + 1
                                        icounts[k] <- shat[i]
                                        if(m > 1)
                                        icountv[k] <- v[i]/inumb[i]
                                        if(m == 1)
                                        icountv[k] <- v[i]
                                }
                        }
                }
            }
            rr <- sorted[, 2]
            ss <- c(icounts[1:(n - 1)], 1)
            ss <- 1/ss
            vv <- icountv * delta
            ss <- (1 - delta) * ss
#w holds the weights in matrix form
            w <- outer(ss, vv)
            w[row(w) > col(w)] <- 0
            w <- w + diag(delta)
            browser()
#output consists of weight matrix and sorted variables
            list(wm = w, r = rr, timex = timex, timez = timez,
                status = status)
}
```

FUNCTIONS FOR RENOVATING CENSORED RESPONSES   231

We now give two functions sur1 and sur2 which are called by the main renovation program and the weight function. These functions are called to deal with ties in censored data; two functions are necessary because ties which affect product-limit estimation may be between uncensored data points, or, censored and uncensored data points. The function sur1 notes the number of tied data points in a sorted vector of positive numbers and holds them in icount.

S-PLUS coding for identifying tied data — sur1()

```
sur1<-function(x) {
        n <- length(x)
        tcount <- 0
        icount <- array(0, n)
        l <- 1
        i <- 1
        for(i in 1:n) {
                tcount <- 0
                for(j in 1:n) {
                        if(x[i] == x[j])
                                tcount <- tcount + 1
                }
                icount[i] <- tcount
                l <- l + tcount
                if(l > n)
                        break
        }
        icount
}
```

The function sur2 records the number of censored data points for each group of tied data and holds them in inumb in the main functions. Note that in such a group of ties, uncensored observations conventionally precede censored observations (Miller, 1976).

S-PLUS coding for censored points in tied data — sur2()

```
sur2<-function(x, y, num) {
        k <- 0
        nn <- length(x)
        z <- (1:nn)
        for(i in 1:nn) {
                m <- x[i]
                numb <- 0
                for(j in 1:m) {
                        k <- k + 1
```

```
                    if(k == num)
                            break
                    if(y[k] == 0)
                            numb <- numb + 1
            }
            if(numb < x[i])
                    z[i] <- x[i] - numb
            if(numb == x[i])
                    z[i] <- 1
        }
        z
}
```

▬ Example 11.5
Using the renovation functions on the Stanford Heart Transplant data

We give the results for applying the renovation functions to the right-censored data from the Stanford Heart Transplant Study (Example 5.2) using a single covariate. Here, the response of interest is log base 10 days survival after transplant for the original 69 patients in the program; age in years is taken as a single covariate and 24 patients were right-censored at the termination of the study after having been admitted to the program at possibly different times. Applying (11.6) to these data results in the fit

$$\log \text{days} = 3.598 - 0.0279 \text{ age},$$

where the slope estimate is an exact zero-crossing of $\Phi(b) = 0$. The input data (listed in Example 5.2) are:

z = heart.log holding log base 10 survival time;
status = heart.d holding the censor indicators;
x = heart.x containing the covariate 'age in years';
num set to 10 iterations.

The output from ren.fit now follows:

```
$covariate:
 [1] 35.2 41.5 54.2 40.4 29.2 28.9 40.3 55.3 36.2 54.3 23.7 45.8
42.8 42.5 [15] 53.0 52.5 19.7 56.9 26.7 54.0 46.4 47.2 45.2 49.0
50.5 53.4 52.5 49.1 [29] 51.3 51.4 54.6 56.4 61.5 43.8 48.0 47.5
26.7 64.5 52.0 42.8 47.8 48.8 [43] 32.7 49.5 48.1 49.3 48.9 46.5
38.9 54.4 36.8 41.6 47.5 48.9 52.9 52.2 [57] 48.0 45.0 33.3 51.0
43.4 46.1 40.6 48.6 45.5 48.6 58.4 49.0 54.1

$oldresponse:
 [1] 0.0000000 0.0000000 0.0000000 0.4771213 1.0791812 1.1139434
```

1.1461280 [8] 1.0000000 1.6434527 1.1760913 2.0413927 1.4771213
1.5910646 1.6627578 [15] 1.3979400 1.4149733 2.3579348 1.3617278
2.2227165 1.4623980 1.6989700 [22] 1.7075702 1.8129134 1.7323938
1.7075702 1.6812412 1.7075702 1.8129134 [29] 1.8195439 1.8325089
1.8061800 1.7993405 1.6720979 2.2068259 2.1038037 [36] 2.1673173
2.7723217 1.7781513 2.1335389 2.4727564 2.3747483 2.4031205 [43]
2.9111576 2.4471580 2.5078559 2.4842998 2.5899496 2.6589648
2.9420081 [50] 2.5301997 3.0437551 2.9232440 2.7701153 2.7411516
2.6424645 2.6981005 [57] 2.8195439 2.9222063 3.2491984 2.7951846
3.0103000 2.9527924 3.1900514 [64] 2.9973864 3.1017471 3.1357685
2.8633229 3.1863912 3.1303338

$response:
[1] 2.6170041 0.0000000 0.0000000 0.4771213 1.0791812 2.9214759
1.1461280 [8] 1.0000000 1.6434527 1.1760913 3.1472126 2.5301118
1.5910646 1.6627578 [15] 1.3979400 1.4149733 2.3579348 1.3617278
3.1732649 1.4623980 1.6989700 [22] 1.7075702 1.8129134 1.7323938
1.7075702 1.6812412 1.7075702 1.8129134 [29] 1.8195439 1.8325089
1.8061800 1.7993405 1.6720979 2.2068259 2.1038037 [36] 2.1673173
3.5315833 1.7781513 2.1335389 2.4727564 3.0113295 2.4031205 [43]
3.4544606 2.4471580 2.5078559 3.0348862 3.0460555 3.1130709
3.3252866 [50] 2.8924783 3.3839251 3.2498942 3.0851478 2.7411516
2.9546899 2.9742361 [57] 3.0915132 2.9222063 3.5249898 2.7951846
3.0103000 2.9527924 3.4187246 [64] 2.9973864 3.3214920 3.2349304
2.8633229 3.2727416 3.1303338

$censorindicator:
 [1] 0 1 1 1 1 0 1 1 1 1 0 0 1 1 1 1 1 1 0 1 1 1 1 1 1 1 1 1 1 1
 1 1 1 1 [36] 1 0 1 1 1 0 1 0 1 1 0 0 0 0 0 0 0 1 0 0 0 1 0 1 1
 1 0 1 0 0 1 0 1

$bjslope:
                X
 3.598403 -0.02789071

$lsslope:
 Intercept   X
 3.598403 -0.02789071

$trace:
          [,1]   [,2]
 [1,] 3.206633 -0.02034199
 [2,] 3.468719 -0.02538055
 [3,] 3.555483 -0.02703624
 [4,] 3.585801 -0.02761285
 [5,] 3.602083 -0.02792311
 [6,] 3.598403 -0.02789071
 [7,] 0.000000 0.00000000
 [8,] 0.000000 0.00000000

```
[9,]  0.000000   0.00000000
[10,] 0.000000   0.00000000

$norm:
        [1]           [2]           [3]           [4]
 0.005038564  0.001655689  0.0005766168  0.0003102564
          [5]  [6] [7] [8] [9]
 0.00003239837   0   0   0   0
```

Notice that, in the case of the Heart Transplant data, the norm of adjacent iterates is sufficiently small (less than 0.0001) after the fifth iteration. This corresponds to an 'exact' zero-crossing. A comparison of corresponding entries in $oldresponse and $response shows that the positions of uncensored responses are not moved within the scatterplot. □

These functions developed in this section may easily have their coding generalised to cater for interval-censored responses (Section 10.8), as follows. In the process of renovation, each censored point is initially positioned at the left endpoint of its own censoring interval, and moved to the right in a direction of increasing response to an estimated position within its own censoring interval.

11.4 The structure of the renovated hat matrix

In linear regression the structure of the hat matrix plays an important part in regression diagnostics. In this section, following the work of Smith and Peiris (1999), we investigate the properties of the renovated hat matrix (Definition 11.2) for regression with censored responses. The structure of the Buckley-James censored regression estimators allows natural links to be established with the structure of ordinary least squares estimators. In particular, the renovated hat matrix may be partitioned in a manner which assists in deciding whether further explanatory variables should be added to the linear model using the 'added variable plot'.

Recall first that the **renovated hat matrix** of Definition 11.2 is written

$$\mathbf{H}^\star = \mathbf{X}(\mathbf{X}^T \mathcal{W} \mathbf{X})^{-1} \mathbf{X}^T \mathcal{W}.$$

Our aim is to uncover a structure of $\mathbf{H}^\star$ based on a partition of $\mathbf{X}$.

Let $\mathbf{X} = (\mathbf{X}_1 : \mathbf{X}_2)$, where $\mathbf{X}_1$ is an $n \times r$ matrix of rank r and $\mathbf{X}_2$ is an $n \times (p-r)$ matrix of rank $p-r$, and let $\mathbf{H}_1^\star$ be defined by

$$\mathbf{H}_1^\star = \mathbf{X}_1(\mathbf{X}_1^T \mathcal{W} \mathbf{X}_1)^{-1} \mathbf{X}_1^T \mathcal{W}.$$

The structural properties of this matrix are given by Theorem 11.2 which may be established by direct substitution:

Theorem 11.2 $\mathbf{H}_1^\star$ *satisfies:*

1. $\mathbf{H}_1^{\star T} \mathcal{W} \mathbf{H}_1^\star = \mathbf{H}_1^{\star T} \mathcal{W}$;
2. $(I - \mathbf{H}_1^\star)^T \mathcal{W}(I - \mathbf{H}_1^\star) = \mathcal{W}(I - \mathbf{H}_1^\star)$.

□

The renovated hat matrix may now be partitioned using Lemma 2.1 of Chatterjee and Hadi (1988) for partitioned matrix inverses. Smith and Peiris (1999) give the details which are structurally outline here:

$$\begin{aligned} \mathbf{H}^\star &= \mathbf{X}(\mathbf{X}^T \mathcal{W} \mathbf{X})^{-1} \mathbf{X}^T \mathcal{W} \\ &= (\mathbf{X}_1 : \mathbf{X}_2) \begin{pmatrix} \mathbf{X}_1^T \mathcal{W} \mathbf{X}_1 & \mathbf{X}_1^T \mathcal{W} \mathbf{X}_2 \\ \mathbf{X}_2^T \mathcal{W} \mathbf{X}_1 & \mathbf{X}_2^T \mathcal{W} \mathbf{X}_2 \end{pmatrix}^{-1} \begin{pmatrix} \mathbf{X}_1^T \mathcal{W} \\ \mathbf{X}_2^T \mathcal{W} \end{pmatrix} \\ &= \mathbf{H}_1^\star + (I - \mathbf{H}_1^\star) \mathbf{X}_2 M \mathbf{X}_2^T \mathcal{W}(I - \mathbf{H}_1^\star), \end{aligned}$$

where

$$\begin{aligned} M &= [(\mathbf{X}_2^T \mathcal{W} \mathbf{X}_2) - (\mathbf{X}_2^T \mathcal{W} \mathbf{X}_1)(\mathbf{X}_1^T \mathcal{W} \mathbf{X}_1)^{-1}(\mathbf{X}_1^T \mathcal{W} \mathbf{X}_2)]^{-1} \\ &= [\mathbf{X}_2^T \mathcal{W}(I - \mathbf{H}_1^\star) \mathbf{X}_2]^{-1}. \end{aligned} \quad (11.7)$$

Therefore,

$$\mathbf{H}^\star = \mathbf{H}_1^\star + (I - \mathbf{H}_1^\star) \mathbf{X}_2 [\mathbf{X}_2^T \mathcal{W}(I - \mathbf{H}_1^\star) \mathbf{X}_2]^{-1} \mathbf{X}_2^T \mathcal{W}(I - \mathbf{H}_1^\star).$$

If we simplify notation using the identifications

$$\mathbf{X}_2^\star = (I - \mathbf{H}_1^\star) \mathbf{X}_2 \quad \text{and} \quad \mathcal{W}^\star = \mathcal{W}(I - \mathbf{H}_1^\star),$$

then $\mathcal{W}^\star$ is idempotent and acts like a 'centered' weights matrix. By Theorem 11.2, $\mathbf{H}^\star$ can be written as

$$\mathbf{H}^\star = \mathbf{H}_1^\star + \mathbf{X}_2^\star (\mathbf{X}_2^{\star T} \mathcal{W}^\star \mathbf{X}_2^\star)^{-1} \mathbf{X}_2^{\star T} \mathcal{W}^\star. \quad (11.8)$$

This establishes the following theorem.

Theorem 11.3 *The renovated hat matrix, $\mathbf{H}^\star$, has the form $\mathbf{H}^\star = \mathbf{H}_1^\star + \mathbf{H}_2^\star$, where $\mathbf{H}_1^\star$ and $\mathbf{H}_2^\star$ are idempotent and $\mathbf{H}_1^\star \mathbf{H}_2^\star = 0 = \mathbf{H}_2^\star \mathbf{H}_1^\star$.* □

The partition of the design matrix $\mathbf{X}$ may be chosen so as to isolate the constant term in the linear model. For this, let $\mathbf{X} = (\mathbf{1} : \mathbf{X}_2)$, where $\mathbf{X}_1 = \mathbf{1}$ is the n-vector of ones. In this case it follows that

$$\mathbf{H}_1^\star = \frac{1}{n} \mathbf{1} \mathbf{1}^T \mathcal{W},$$

so that it is easy to establish that $\mathbf{H}^\star \mathbf{1} = \mathbf{1}$ using Theorem 11.3.

11.5 Added variable plots from Buckley-James estimators

Many authors have contributed to the recent development of the **added variable plot** (also called the **partial regression plot**) in the context of ordinary linear regression diagnostics (Mosteller and Tukey, 1977; Cook and Weisberg, 1982; Davidson and Snell, 1991). As the name suggests, this plot is a graphical device which allows us to investigate the effect of adding a covariate to the linear model. In another sense, the plot may be used to assess whether there is any relationship between the scatterplot of $\mathbf{Y}$ and a new covariate, and, the scatterplot for $\mathbf{Y}$ adjusted for covariates already in the model and a new covariate. Also, this graphical device is used to find non-linearity and to assist in the detection of influential points (Cook, 1996).

In this section we extend this graphical technique to the Buckley-James regression methodology so as to produce added variable plots for regression with censored data. In parallel with ordinary least squares, the spirit of such a plot is to look for a linear effect on a scatterplot of one set of specially renovated residuals against another (Definition 11.3).

Again reconsider the standard censored linear regression model $\mathbf{Y} = \mathbf{X}_1\boldsymbol{\beta}_1 + \boldsymbol{\epsilon}_1$ of (11.6), where $\mathbf{X}_1$ is of full rank with a possible first column of ones. Now we consider the introduction of an additional explanatory variable, $\mathbf{X}_2$, represented in the model

$$\mathbf{Y} = \mathbf{X}_1\boldsymbol{\beta}_1 + \mathbf{X}_2\boldsymbol{\beta}_2 + \boldsymbol{\epsilon}_2 \qquad (11.9)$$

as an $n \times 1$ vector with associated parameter β_2. The error distribution represented in (11.9) by $\boldsymbol{\epsilon}_2$ has mean 0 and finite variance.

Consider first the Buckley-James estimation of $\boldsymbol{\beta} = (\boldsymbol{\beta}_1, \boldsymbol{\beta}_2)^T$.

Theorem 11.4 *Under the conditions of an exact zero-crossing for $\Phi_n(\mathbf{b}) = 0$ in (10.16), the Buckley-James estimator $\widehat{\beta_2}$ of β_2 in the model $\mathbf{Y} = \mathbf{X}_1\boldsymbol{\beta}_1 + \mathbf{X}_2\boldsymbol{\beta}_2 + \boldsymbol{\epsilon}_2$ is given by*

$$\widehat{\beta}_2 = [\mathbf{X}_2^T \mathcal{W}(I - \mathbf{H}_1^\star)\mathbf{X}_2]^{-1}[\mathbf{X}_2^T \mathcal{W}(I - \mathbf{H}_1^\star)]\mathbf{Y}^\star,$$

where $\mathcal{W}$ is the renovation weight matrix for fitting the full model containing both $\mathbf{X}_1$ and $\mathbf{X}_2$ with $\mathbf{H}_1^\star = \mathbf{X}_1(\mathbf{X}_1^T \mathcal{W} \mathbf{X}_1)^{-1}\mathbf{X}_1^T \mathcal{W}$.

Proof: Let $\widehat{\beta}_i$ be the Buckley-James estimate of β_i, $i = 1, 2$ under the conditions of an exact zero-crossing of (10.16) with $\mathbf{X} = (\mathbf{X}_1 : \mathbf{X}_2)$.

$$\begin{pmatrix} \widehat{\beta}_1 \\ \widehat{\beta}_2 \end{pmatrix} = \left[(\mathbf{X}_1 : \mathbf{X}_2)^T \mathcal{W}(\mathbf{X}_1 : \mathbf{X}_2)\right]^{-1}(\mathbf{X}_1 : \mathbf{X}_2)^T \mathcal{W}\mathbf{Y}^\star$$

$$= \begin{pmatrix} \mathbf{X}_1^T \mathcal{W} \mathbf{X}_1 & \mathbf{X}_1^T \mathcal{W} \mathbf{X}_2 \\ \mathbf{X}_2^T \mathcal{W} \mathbf{X}_1 & \mathbf{X}_2^T \mathcal{W} \mathbf{X}_2 \end{pmatrix}^{-1} \begin{pmatrix} \mathbf{X}_1^T \mathcal{W} \mathbf{Y}^\star \\ \mathbf{X}_2^T \mathcal{W} \mathbf{Y}^\star \end{pmatrix}$$

and the result now follows by simplifying the inverted matrix and using Theorem 11.2. □

Theorem 11.4 plays a role in the structure of residuals of the added variable plot. We give two definitions of the added variable plot in the context of censored linear regression. In Definition 11.3, we specify the quantities involved in the plot as matrices and vectors (as a working definition) and in Definition 11.4, we indicate that our censored version of the added variable plot is identical in spirit to the uncensored case.

Definition 11.3 *By an* **added variable plot** *for* $\mathbf{X}_2$ *in the censored linear regression model* $\mathbf{Y} = \mathbf{X}_1\boldsymbol{\beta}_1 + \mathbf{X}_2\boldsymbol{\beta}_2 + \epsilon$, *we mean a plot of* $(I - \mathbf{H}_1^\star)\mathbf{Y}^\star$ *against* $(I - \mathbf{H}_1^\star)\mathbf{X}_2$, *where* $\mathbf{Y}^\star$ *are the renovated responses based on the full model containing both* $\mathbf{X}_1$ *and* $\mathbf{X}_2$, *with corresponding renovation weight matrix* $\mathcal{W}$ *and* $\mathbf{H}_1^\star = \mathbf{X}_1(\mathbf{X}_1^T\mathcal{W}\mathbf{X}_1)^{-1}\mathbf{X}_1^T\mathcal{W}$. □

The motivation of Definition 11.3, and the 'residual' character of an added-variable plot, may be understood through the renovation process. We begin with the full model (11.9):

$$\mathbf{Y} = \mathbf{X}_1\boldsymbol{\beta}_1 + \mathbf{X}_2\boldsymbol{\beta}_2 + \boldsymbol{\epsilon}_2.$$

Suppose that $\boldsymbol{\beta} = (\boldsymbol{\beta}_1, \boldsymbol{\beta}_2)^T$ has renovation slope estimator

$$\widehat{\boldsymbol{\beta}} = (\mathbf{X}^T\mathcal{W}\mathbf{X})^{-1}\mathbf{X}^T\mathcal{W}\mathbf{Y}^\star,$$

where

$$\mathbf{Y}^\star = \mathbf{X}\widehat{\boldsymbol{\beta}} + \mathcal{W}(\mathbf{Z} - \mathbf{X}\widehat{\boldsymbol{\beta}}) \quad (11.10)$$

denotes the renovated responses under the renovation weight matrix $\mathcal{W}$ and $\mathbf{X} = (\mathbf{X}_1 : \mathbf{X}_2)$. Further, let $\widehat{\boldsymbol{\beta}_1} = (\mathbf{X}_1^T\mathcal{W}\mathbf{X}_1)^{-1}\mathbf{X}_1^T\mathcal{W}\mathbf{Z}$ and, following the notation of Cook (1996), let $\epsilon_{y|1}^\star = \mathbf{Y}^\star - \mathbf{X}_1\widehat{\boldsymbol{\beta}_1}$ denote the corresponding renovated residuals associated with $\widehat{\boldsymbol{\beta}_1}$ in the fit with $\mathbf{X}_1$.
Finally, let

$$\widehat{\boldsymbol{\beta}_{2|1}} = (\mathbf{X}_1^T\mathcal{W}\mathbf{X}_1)^{-1}\mathbf{X}_1^T\mathcal{W}\mathbf{X}_2.$$

If we consider a renovated form of the left-side of (11.9) and replace $\mathbf{Y}$ by $\mathbf{Y}^\star$, then $\mathbf{Y}^\star$ still follows a linear model (see, for example, Theorem 10.1 in this regard, where essentially $E(\mathbf{Y}^\star) = E(\mathbf{Y})$ when the true regression parameters are known). Without loss of generality, on cross-multiplying each side of (11.9) by $(I - \mathbf{H}_1^\star)$, we obtain

$$(I - \mathbf{H}_1^\star)\mathbf{Y}^\star = (I - \mathbf{H}_1^\star)\mathbf{X}_1\boldsymbol{\beta}_1 + (I - \mathbf{H}_1^\star)\mathbf{X}_2\boldsymbol{\beta}_2 + (I - \mathbf{H}_1^\star)\boldsymbol{\epsilon}_2. \quad (11.11)$$

Since $(I - \mathbf{H}_1^\star)\mathbf{X}_1 = 0$, we may write

$$(I - \mathbf{H}_1^\star)\mathbf{Y}^\star = (I - \mathbf{H}_1^\star)\mathbf{X}_2\boldsymbol{\beta}_2 + (I - \mathbf{H}_1^\star)\boldsymbol{\epsilon}_2. \quad (11.12)$$

Note that the left-side of (11.12) is $\epsilon_{y|1}^\star$, the renovated residuals in $\mathbf{Y}$ after fitting $\mathbf{X}_1$. The coefficient of β_2 in the first term on the right-side of (11.12) is $\epsilon_{2|1}^\star$, the renovated residuals in $\mathbf{X}_2$ after fitting $\mathbf{X}_1$. Note particularly that these renovations are each accomplished through $\mathcal{W}$, which has the effect of maintaining the same residual ranking for all sets of residuals being compared in the plots. That is, the residuals being plotted have order induced by the residuals from the full model Buckley-James renovation of $\mathbf{Y}$ against $\mathbf{X} = (\mathbf{X}_1 : \mathbf{X}_2)$.

Now (11.12) can be written as

$$\epsilon^\star_{y|1} = \epsilon^\star_{2|1}\beta_2 + \epsilon_3 \qquad (11.13)$$

where ϵ_3 has mean 0 and finite variance if we regard $\mathcal{W}$ as fixed. (See Currie, 1996, for comments regarding the constancy of $\mathcal{W}$ during the Buckley-James iterative process.) This allows us to define the added variable plot in terms of residuals.

Definition 11.4 *By an* **added variable plot for $\mathbf{X}_2$** *in the censored linear model* $\mathbf{Y} = \mathbf{X}_1\boldsymbol{\beta}_1 + \mathbf{X}_2\boldsymbol{\beta}_2 + \boldsymbol{\epsilon}$, *we mean a plot of* $\epsilon^\star_{y|1}$, *the renovated residuals in* $\mathbf{Y}$ *after fitting* $\mathbf{X}_1$, *against* $\epsilon^\star_{2|1}$, *the renovated residuals in* $\mathbf{X}_2$ *after fitting* $\mathbf{X}_1$. □

This definition allows us to develop several properties for added variable plots. The least squares estimator $\widehat{\beta'_2}$ (say) for β_2 in model (11.13) is

$$\widehat{\beta'_2} = (\epsilon^{\star T}_{2|1}\epsilon^\star_{2|1})^{-1}\epsilon^{\star T}_{2|1}\epsilon^\star_{y|1} \qquad (11.14)$$

and the renovation slope estimator, $\widehat{\beta''_2}$ (say) for β_2 in model (11.13) is

$$\widehat{\beta''_2} = (\epsilon^{\star T}_{2|1}\mathcal{W}\epsilon^\star_{2|1})^{-1}\epsilon^{\star T}_{2|1}\mathcal{W}\epsilon^\star_{y|1}. \qquad (11.15)$$

Using an argument involving traces, it may be established that

$$\widehat{\beta''_2} = \widehat{\beta'_2} = \widehat{\beta_2}.$$

This means that a scatterplot of $\epsilon^\star_{y|1}$ against $\epsilon^\star_{2|1}$ showing a linear effect would have a regression line with least squares slope equal to the Buckley-James slope estimate $\widehat{\beta_2}$ or the renovation slope estimate $\widehat{\beta''_2}$.

Suppose that we use $\widehat{\beta''_2}$ to renovate the added variable plot of $\epsilon^\star_{y|1}$ against $\epsilon^\star_{2|1}$. For this purpose we can use the standard renovation equation

$$\epsilon^\star_{y|1}{}^\star = \epsilon^\star_{2|1}\widehat{\beta''_2} + \mathcal{W}(\epsilon^\star_{y|1} - \epsilon^\star_{2|1}\widehat{\beta''_2}).$$

If we define $\mathbf{H}^\star_{2|1} = \epsilon^\star_{2|1}(\epsilon^{\star T}_{2|1}\mathcal{W}\epsilon^\star_{2|1})^{-1}\epsilon^{\star T}_{2|1}\mathcal{W}$, then the renovated residuals from fitting model (11.13) may be written

$$\begin{aligned}(I - \mathbf{H}^\star_{2|1})\epsilon^\star_{y|1}{}^\star &= \mathcal{W}(I - \mathbf{H}^\star_{2|1})\epsilon^\star_{y|1} \\ &= \mathcal{W}(I - \mathbf{H}^\star_{2|1})(I - \mathbf{H}^\star_1)\mathbf{Y}^\star.\end{aligned}$$

Using Theorem 11.3, the renovated hat matrix of model (11.9) can be written as

$$\mathbf{H}^\star = \mathbf{H}^\star_1 + (I - \mathbf{H}^\star_1)\mathbf{X}_2[\mathbf{X}^T_2\mathcal{W}(I - \mathbf{H}^\star_1)\mathbf{X}_2]^{-1}\mathbf{X}^T_2\mathcal{W}(I - \mathbf{H}^\star_1).$$

Hence, by Theorem 11.2,

$$\begin{aligned}(I - \mathbf{H}^\star)\mathbf{Y}^\star &= \mathcal{W}(I - \mathbf{H}^\star)\mathbf{Y}^\star \\ &= \mathcal{W}[I - \epsilon^\star_{2|1}(\epsilon^{\star T}_{2|1}\mathcal{W}\epsilon^\star_{2|1})^{-1}\epsilon^{\star T}_{2|1}\mathcal{W}](I - \mathbf{H}^\star_1)\mathbf{Y}^\star\end{aligned}$$

$$\begin{aligned} &= \mathcal{W}(I - \mathbf{H}^\star_{2|1})(I - \mathbf{H}^\star_1)\mathbf{Y}^\star \\ &= (I - \mathbf{H}^\star_{2|1})\boldsymbol{\epsilon}^\star_{y|1}{}^\star. \end{aligned}$$

This means that the renovated residuals from fitting the full model (11.9) are the same as the residuals from model (11.13) which would arise if the added variable plot were renovated using $\mathcal{W}$ and $\widehat{\beta''_2}$. This establishes the following theorem.

Theorem 11.5 *The added variable plot for $\mathbf{X}_2$ in the censored linear full model $\mathbf{Y} = \mathbf{X}_1\boldsymbol{\beta}_1 + \mathbf{X}_2\boldsymbol{\beta}_2 + \boldsymbol{\epsilon}$ has the same least squares slope estimator as the Buckley-James estimator $\widehat{\beta_2}$ of β_2 in the full model and, if the added variable plot is renovated using the full model weight matrix $\mathcal{W}$, it has the same renovated residuals as the full model.* □

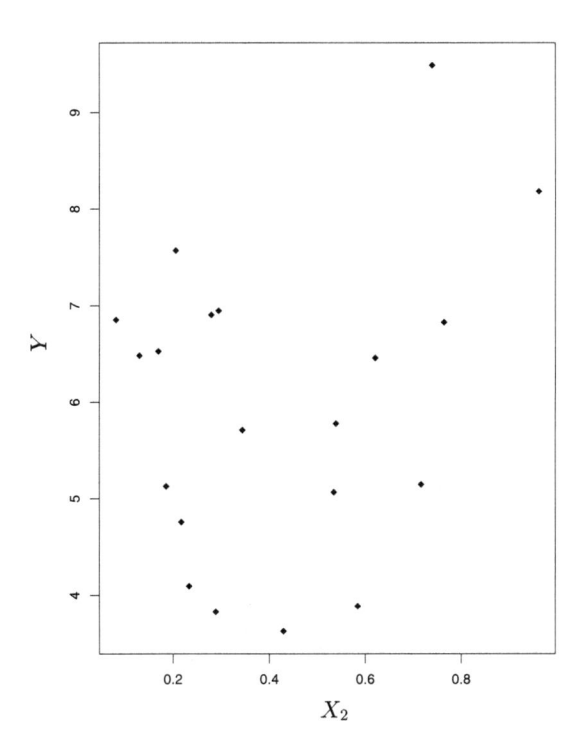

Figure 11.4 *A scatterplot of Y versus X_2 for simulated data with uniform covariates and normal responses.*

We consider two examples to illustrate the scope of added variable plots as a diagnostic tool in censored linear regression: first, as an outlier detector

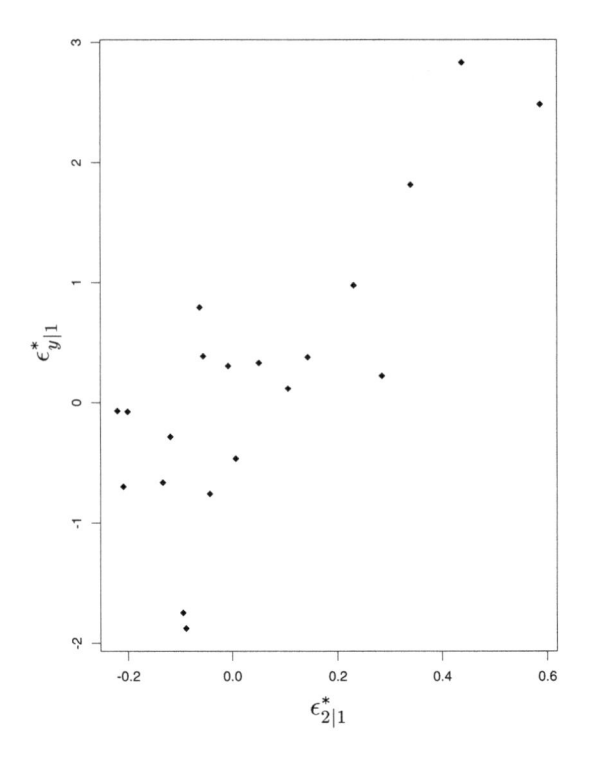

Figure 11.5 *An added variable plot for X_2 for simulated data with uniform covariates and normal responses. Possible outliers in lower left corner. Confirmatory linear trend.*

in a linear model with two synthetically generated covariates and normally distributed residuals; second, as a quadratic curvature check for the Stanford Heart Transplant data of Example 5.2.

■ **Example 11.6**
Constructing an added variable plot and its use in outlier detection
Following the example of Smith and Peiris (1999), we use standard notation for the linear model with normally generated residuals. Instead of recording responses, **Y**, the observed data are generated by

$$(z_i, \delta_i, \mathbf{x}_i), \quad i = 1, 2, \ldots, 20,$$

where δ_i are independent and identically distributed $Binomial(1, 0.7)$ variables, giving a 30% chance of censoring; $\mathbf{x}_i^T = (1, x1_i, x2_i)$, where $x1_i$ and $x2_i$ are independent and identically distributed uniform random variables on $[0, 1]$; and

$$Z_i = 1 + 5X1_i + 3X2_i + \epsilon_i,$$

ADDED VARIABLE PLOTS FROM BUCKLEY-JAMES ESTIMATORS 241

where $\epsilon_i \sim N(0,1)$. This simulation produced realised data with 25% censoring.

Note that in terms of the notation of this chapter, we may easily identify the structure of the partition $\mathbf{X} = (\mathbf{X}_1 : \mathbf{X}_2)$ in this simulation. Here, $\mathbf{X}_1$ is a 20×2 matrix with the first column of 1's and the second column containing the values $x1_i$, $i = 1, 2, \ldots, 20$. Further, $\mathbf{X}_2$ is a vector containing the entries $x2_i$, $i = 1, 2, \ldots, 20$ as observations on the explanatory variable X_2.

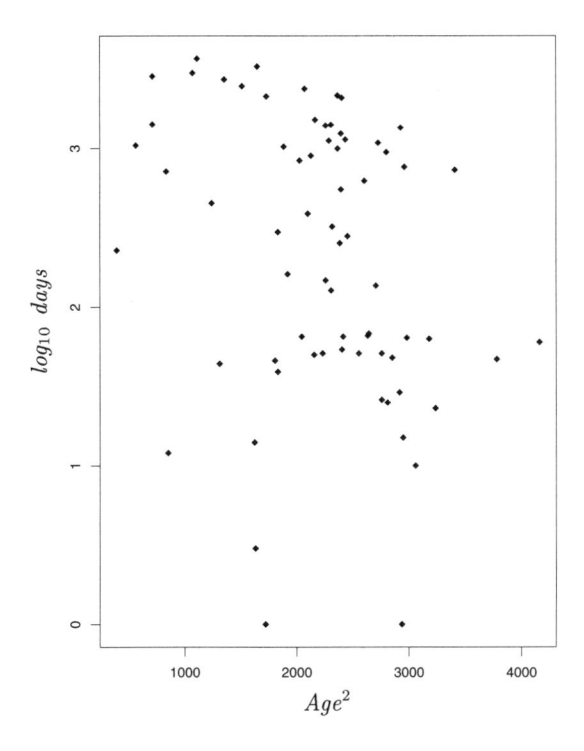

Figure 11.6 *A scatterplot of $\log_{10}$ days versus Age^2 for the Stanford Heart Transplant Ddata.*

To create an added variable plot for X_2 (Figure 11.5), we first create the renovated scatterplot of $\mathbf{Z}$ on both X_1 and X_2 and then we investigate whether the term X_2 is really necessary to explain the variability in the response. Figure 11.4 is the scatterplot of the observed response against X_2 showing some scatter and a sense of linear trend. The added variable plot for X_2 (Figure 11.5) shows a heightened sense of linearity, indicating that X_2 is likely to have a significant regression coefficient and should not be omitted from the model.

The added variable plot also indicates the presence of two possible outliers in

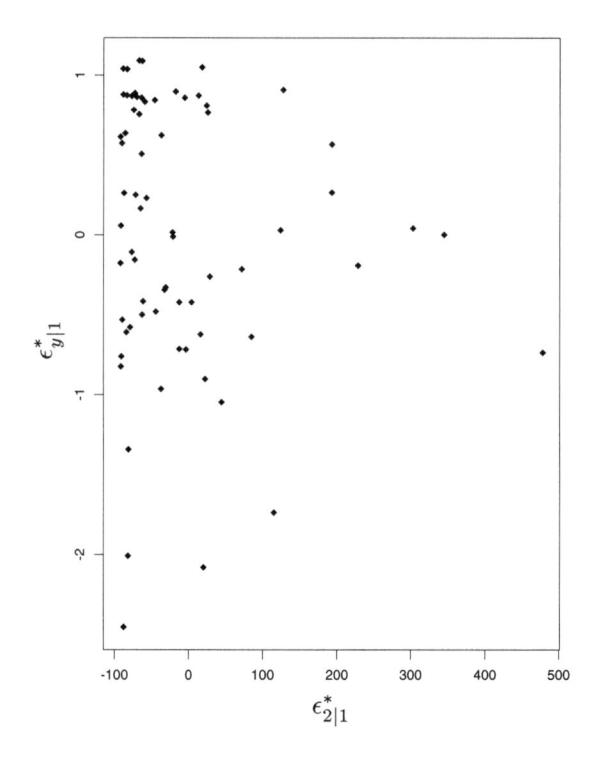

Figure 11.7 *An added variable plot for Age^2 for the Stanford Heart Transplant data. The plot still shows scatter and therefore does not confirm a quadratic trend in Age for these data.*

the lower left corner of the plot. Removing one of these outliers does not effect the Buckley-James slope significantly (an example of a **masking effect**), but these two points jointly influence the Buckley-James slope. The Buckley-James estimators of the parameters in the model $Y = \beta_0 + \beta_1 X1 + \beta_2 X2 + \epsilon$, obtained from an exact zero-crossing of (10.16), may be read from the fitted models:

$$y = 1.024 + 5.084x1 + 4.237x2,$$

including the possible outliers;

$$y = 1.218 + 5.489x1 + 3.57x2$$

with the possible outliers excluded. The Buckley-James fit shows a considerable change after the outliers are removed, confirming that the points are having an undue effect on the fitting process and in many contexts, would be deleted. □

■ Example 11.7
Checking for a quadratic curvature for *Age* in the Stanford Heart Transplant data using an added variable plot

It is difficult to tell how the variable *Age* enters into the censored regression model for the Stanford Heart Transplant data. It is generally agreed that *Age* on its own is a significant inclusion in the model that has $\log_{10}(days)$ as responses. A scatterplot of $\log_{10}(days)$ against *Age* shows considerable scatter, so that the *Age* effect is only just significant.

Here we effectively check as to whether there is any curvature in this scatterplot. Indeed, it is only worth adding the variable Age^2 into the model with *Age* present if curvature is present. The curvature can be checked through an added variable plot for Age^2.

Figure 11.6 and Figure 11.7 are the scatterplots of $log_{10}(days)$ of survival versus Age^2 and the added variable plot for Age^2, respectively. Clearly, the added variable plot for Age^2 does not exhibit linearity, confirming that Age^2 is not a significant factor in determining heart transplant survival — an effect recorded previously by Miller and Halpern (1982). □

11.6 Exercises

11.1. Show that if the renovation slope estimate $\widehat{\beta}$ satisfies (11.2) and (11.3), then
$$\widehat{\beta_n} = \widehat{\beta} \Leftrightarrow \Phi(\widehat{\beta_n}) = 0.$$

11.2. Show that the renovated hat matrix $\mathbf{H}^*$, resulting from a Buckley-James fit to a scatterplot, satisfies $(\mathbf{H}^*)^2 = \mathbf{H}^*$ and $trace(\mathbf{H}^*) = p$ when $\mathbf{X}$ is $n \times p$ of full rank p.

11.3. Establish the result of Theorem 11.2 that $\mathbf{H}_1^*$ satisfies
$$\mathbf{H}_1^{*T} \mathcal{W} \mathbf{H}_1^* = \mathbf{H}_1^{*T} \mathcal{W}$$
and
$$(I - \mathbf{H}_1^*)^T \mathcal{W}(I - \mathbf{H}_1^*) = \mathcal{W}(I - \mathbf{H}_1^*).$$

11.4. Show that $\mathcal{W}^* = \mathcal{W}(I - \mathbf{H}_1^*)$ is idempotent.

11.5. Given the structure of $\mathbf{H}_1^*$ and $\mathbf{H}_2^*$ in $\mathbf{H}^* = \mathbf{H}_1^* + \mathbf{H}_2^*$, show that $\mathbf{H}_1^*$ and $\mathbf{H}_2^*$ are idempotent and that $\mathbf{H}_1^* \mathbf{H}_2^* = 0 = \mathbf{H}_2^* \mathbf{H}_1^*$.

11.6. Use Theorem 11.3 to show that if $\mathbf{X} = (\mathbf{1} : \mathbf{X}_2)$ is a design matrix with first column of ones, then the renovated hat matrix satisfies $\mathbf{H}^* \mathbf{1} = \mathbf{1}$.

11.7. Collett (1994) reports data for the time to recurrence of an ulcer after an initial ulcer has been diagnosed, treated and healed. The data for 42 patients are listed with the intervals (in our notation) I_i for $i = 1, 2, 3, \ldots, 42$ appended.

Age	Treatment	Time of last visit	Result	Censoring interval
48	B	7	2	$\{7\}$
73	B	12	1	$[12, \infty]$
54	B	12	1	$[12, \infty]$
58	B	12	1	$[12, \infty]$
56	A	12	1	$[12, \infty]$
49	A	12	1	$[12, \infty]$
71	B	12	1	$[12, \infty]$
41	A	12	1	$[12, \infty]$
23	B	12	1	$[12, \infty]$
37	B	5	2	$\{5\}$
38	B	12	1	$[12, \infty]$
76	B	12	1	$[12, \infty]$
38	A	12	1	$[12, \infty]$
27	A	6	2	$[0, 6]$
47	B	6	2	$[0, 6]$
54	A	6	1	$[6, \infty]$
38	B	10	2	$\{10\}$
27	B	7	2	$\{7\}$
58	A	12	1	$[12, \infty]$
75	B	12	1	$[12, \infty]$
25	A	12	1	$[12, \infty]$
58	A	12	1	$[12, \infty]$
63	B	12	1	$[12, \infty]$
41	A	12	1	$[12, \infty]$
47	B	12	1	$[12, \infty]$
58	A	3	2	$\{3\}$
74	A	2	2	$\{2\}$
75	A	6	1	$[6, \infty]$
72	A	12	1	$[12, \infty]$
59	B	12	2	$[6, 12]$
52	B	12	1	$[12, \infty]$
75	B	12	2	$[6, 12]$
76	A	12	1	$[12, \infty]$
34	A	6	1	$[6, \infty]$
36	B	12	1	$[12, \infty]$
59	B	12	1	$[12, \infty]$
44	A	12	2	$[6, 12]$
28	B	12	1	$[12, \infty]$
62	B	12	1	$[12, \infty]$
23	A	12	1	$[12, \infty]$
49	B	12	1	$[12, \infty]$
61	A	12	1	$[12, \infty]$
33	B	12	1	$[12, \infty]$

The data arose from six monthly clinic visits where endoscopies were performed: a positive test result (result = 2) indicates that the remission ceased in the time interval since the last scheduled visit, providing interval-censored data of the form [0, 6] or [6, 12]; a negative test result (result =1) indicates that the patient is still in remission, providing interval-censored data of the form $[6, \infty]$, or $[12, \infty]$. Further, endoscopies were performed at in-between times for concerned patients presenting with symptoms. Such patients with positive test results provide 'exact' observed remission times. Also recorded was the patient's age (years) at entry into the program and treatment type (A or B). Thus, the data are a combination of traditional left-, right- and interval-censoring.

Fit a Buckley-James regression model to these data with response given by $\log_e$(remission time +1), being the appropriate scale of measurement for an accelerated lifetime model. Use Treatment and/or Age as a covariate(s) and discuss the results.

Does it appear that the 74-year-old patient with a 2-month recurrence time is influential in the positioning of the Buckley-James line?

11.8. For the Stanford Heart Transplant data of Example 5.2, use an added variable plot to assess the inclusion of the 'mismatch score', T5, into the log-linear model which already contains 'Age'. (Most studies have found this variable not to be significant.)

References

Aalen, O.O. (1978). Nonparametric inference for a family of counting processes. *Annals of Statistics*, **6**, 534–45.

Barlow, R.E. and Proschan, F. (1975). *Statistical Theory of Reliability and Life Testing.* Holt, Rinehart & Winston: New York.

Baxter, L.A. and Li, L. (1996). Nonparametric estimation of limiting availability. *Lifetime Data Analysis*, **2**, 391–403.

Belsley, D.A., Kuh, E. and Welsch, R.E. (1980). *Regression Diagnostics.* Wiley: New York.

Blom, G. (1958). *Statistical Estimates and Transformed Beta-Variables.* Wiley: New York.

Box, G.E.P. and Cox, D.R. (1964). An analysis of transformations (with discussion). *Journal of the Royal Statistical Society*, B, **26**, 211–52.

Breslow, N. E. (1974). Covariance analysis of censored survival data. *Biometrics*, **30**, 89–100.

Brick, M.J., Michael, J.R. and Morganstein, D. (1989). Using statistical thinking to solve maintenance problems. *Quality Progress*, May.

Buckley, J.J. and James, I.R. (1979). Linear regression with censored data. *Biometrika*, **66**, 429–36.

Cameron, R.J. (1985). *Year Book Australia 1985.* Australian Bureau of Statistics: Canberra.

Chambers, J.M., Cleveland, W.S., Kleiner, B. and Tukey, P.A. (1983). *Graphical Methods for Data Analysis.* Wadsworth: California.

Chang, M.N. and Yang, G.L. (1987). Strong consistency of a nonparametric estimator of the survival function with doubly censored data. *Annals of Statistics*, **15**, 1536–47.

Chatfield, C. (1983). *Statistics for Technology — A Course in Applied Statistics*, 3rd ed. Chapman & Hall: London.

Chatterjee, S. and Hadi, A.S. (1988). *Sensitivity Analysis in Linear Regression.* Wiley: New York.

Cohen, A.C. (1991). *Truncated and Censored Samples.* Marcel Dekker: New York.

Collett, D. (1994). *Modelling Survival Data in Medical Research.* Chapman & Hall: London.

Cook, R.D. (1996). Added-variable plots and curvature in linear regression. *Technometrics*, **38**, 275–8.

REFERENCES

Cook, R.D., and Weisberg, S. (1982). *Residuals and Influence in Regression.* Chapman and Hall: London.

Cox, D.R. (1972). Regression models and life tables. *Journal of the Royal Statistical Society B*, **34**, 187–202.

Cox, D.R. and Hinkley, D.V. (1974). *Theoretical Statistics.* Chapman & Hall: London.

Cox, D.R. and Snell, E.J. (1968). A general definition of residuals (with discussion). *Journal of the Royal Statistical Society*, B, **30**, 248–75.

Crowder, M.J., Kimber, A.C., Smith, R.L. and Sweeting, T.J. (1991). *Statistical Analysis of Reliability Data.* Chapman and Hall: London.

Crowley, J. and Hu, M. (1977). The covariance analysis of heart transplant data. *Journal of the American Statistical Association*, **72**, 27–36.

Currie, I.E. (1996). A note on Buckley-James estimators for censored data. *Biometrika*, **83**, 912–16.

Davidson, A.C., and Snell, E.J. (1991) Residuals and Diagnostics, in *Statistical Theory and Modelling*, Eds. D.V. Hinkley, N. Reid, and E.S. Snell, Chapman & Hall: London, pp. 83–105.

De Stavola, B.L. and Christensen, E. (1996). Multilevel models for longitudinal variables prognostic for survival. *Lifetime Data Analysis*, **2**, 329–47.

Dinse, G.E. (1985). An alternative to Efron's redistribution-of-mass construction of the Kaplan-Meier estimator. *The American Statistician*, **39**, 299–300.

Efron, B. (1967). The two sample problem with censored data. *Proceedings of the 5th Berkeley Symposium*, **4**, 831–53.

Feigelson, E.D. and Babu, G.J. (1996). *Astrostatistics.* Chapman & Hall: London.

Finkelstein, D.M. (1986). A proportional hazards model for interval-censored failure time data. *Biometrics,* **42**, 845–54.

Finkelstein, D.M. and Wolfe, R.A. (1985). A semiparametric model for regression analysis of interval-censored failure time data. *Biometrics*, **41**, 933–45.

Gehan, E.A. (1965). A generalised Wilcoxon test for comparing arbitrarily singly-censored samples. *Biometrika*, **52**, 203–23.

Gentleman, R. and Crowley, J.J. (1991). Graphical methods for censored data. *Journal of the American Statistical Association*, **86**, 678–83.

Gentleman, R. and Crowley, J.J. (1992). A graphical approach to the analysis of censored data. *Breast Cancer Research and Treatment*, **22**, 229–40.

Greenwood, M. (1926). The natural duration of cancer. *Reports on Public Health and Medical Subjects*, **33**, 1–26, His Majesty's Stationery Office: London.

Gross, A.J. and Clark, V.A. (1975). *Survival Distributions: Reliability Applications in the Biomedical Sciences.* Wiley: New York.

Gulati, S. and Padgett, W.J. (1996). Families of smooth confidence bands for the survival function under the general random censorship model. *Lifetime Data Analysis*, **2**, 349–62.

Heller, G. and Simonoff, J.S. (1990). A comparison of estimators for regression with a censored response variable. *Biometrika*, **77**, 515–20.

Heller, G. and Simonoff, J.S. (1992). Prediction in censored survival data: a comparison of proportional hazards and linear regression models. *Biometrics*, **48**, 101–15.

Hillis, S.L. (1993). A comparison of three Buckley-James variance estimators. *Communications in Statistics*, B, **22**, 955–73.

Hillis, S.L. (1994). A heuristic generalisation of Smith's Buckley-James variance estimator. *Communications in Statistics*, A, **23**, 813–81.

Hoaglin, D.C., Mosteller, F. and Tukey, J.W. (1983). *Understanding Robust and Exploratory Data Analysis.* Wiley: New York.

Hoel, D.G. and Walburg, H.E. Jr. (1972). Statistical analysis of survival experiments. *Journal of the National Cancer Institute*, **49**, 361–2.

Hogg, R.V. and Craig, A.T. (1978). *Introduction to Mathematical Statistics*, 4th edn. Macmillam: New York.

Horner, R.D. (1987). Age at onset of Alzheimer's disease: clue to the relative importance of etiologic factors? *American Journal of Epidemiology*, **126**, 409–14.

James, I.R. and Smith, P.J. (1984). Consistency results for linear regression with censored data. *Annals of Statistics*, **12**, 590–600.

Kalbfleisch, J.D. and Prentice, R.L. (1973). Marginal likelihoods based on Cox's regression and life model. *Biometrika*, **60**, 267–78.

Kalbfleisch, J.D. and Prentice, R.L. (1980). *The Statistical Analysis of Failure Time Data.* Wiley: New York.

Klein, J.P. and Moeschberger, M.L. (1997). *Survival Analysis: Techniques for Censored and Truncated Data.* Springer-Verlag: New York.

Kleinbaum, D.G., Kupper, L.L. and Muller, K.E. (1988). *Applied Regression Analysis and Other Multivariable Methods*, 2nd ed. Duxbury Press: California.

Koul, H., Susarla, V. and Van Ryzin, J. (1981). Regression analysis with randomly right-censored data. *Annals of Statistics*, **9**, 1276–88.

Lai, T.L. and Ying, Z. (1991). Large sample theory of a modified Buckley-James estimator for regression analysis with censored data. *Annals of Statistics*, **19**, 1370–1402.

Lawless, J.F. (1982). *Statistical Models and Methods for Lifetime Data.* Wiley: New York.

Lin, J.S. and Wei, L.J. (1992a). Linear regression analysis for multivariate failure time observations. *Journal of the American Statistical Association*, **87**, 1091–7.

Lin, J.S. and Wei, L.J. (1992). Regression analysis based on Buckley-James estimating equation. *Biometrics*, **48**, 679–81.

Loughin, T.M. and Koehler, K.J. (1997). Bootstrapping regression parameters in multivariate survival analysis. *Lifetime Data Analysis*, **3**, 157–77.

Mann, N.R., and Fertig, K.W. (1973). Tables for obtaining confidence bounds and tolerance bounds based on best linear invariant estimates of parameters of the extreme value distribution. *Technometrics*, **15**, 87–101.

Mantel, N. and Haenszel, W. (1959). Statistical aspects of the analysis of data from retrospective studies of disease. *Journal of the National Cancer Institute*, **22**, 719–48.

Mauro, D.W. (1983). A short note on the consistency of Kaplan-Meier least squares estimators. *Biometrika*, **70**, 534–35.

McCool, J.I. (1980). Confidence limits for Weibull regression with censored data. *IEEE Transaction in Reliability*, **R29**, 145–50.

Mendenhall, W. and Sincich, J. (1992). *Statistics for Engineering and the Sciences*, 3rd ed. Maxwell Macmillan: Singapore.

Meier, P. (1975). Estimation of a distribution function from incomplete observations. *Perspectives in Probability and Statistics*, Ed. J. Gani, 67–87. Academic Press: New York.

Miller, R.G. (1976). Least squares regression with censored data. *Biometrika*, **63**, 449–64.

Miller, R.G. (1981). *Survival Analysis*. Wiley: New York.

Miller, R.G. and Halpern, J. (1982). Regression with censored data. *Biometrika*, **69**, 521–31.

Mosteller, F. and Tukey, J.W. (1977). *Data Analysis and Regression*, Addison-Wesley: Reading, Massachusets.

Myers, R.H. (1990). *Classical and Modern Regression with Applications*, 2nd ed. PWS-Kent: Boston.

Myhre, J.M. (1983). A decreasing failure rate, mixed exponential model applied to reliability. *Reliability in the Acquisitions Process*. D.J. DePriest and R.L. Launer (Eds). Marcel Dekker: New York.

Nair, V.N. (1984). Confidence bands for survival functions with censored data: a comparative study. *Technometrics*, **26**, 265–75.

Nelson, W. (1972). Theory and applications of hazard plotting for censored failure data. *Technometrics*, **14**, 945–65.

Nelson, W. (1972). Graphical analysis of accelerated life test data with the inverse power law model. *IEEE Transactions on Reliability*, **R21**, 2–11.

Parzen, E. (1979). Nonparametric statistical data modelling. *Journal of the American Statistical Association*, **74**, 105–31.

Peto, R. (1973). Experimental survival curves for interval-censored data. *Applied Statistics*, **22**, 86–91.

Potosky, A.L., Riley, G.F., Lubitz, J.D., Mentnech, R.M. and Kessler, L.G. - (1993). Potential for cancer related health services research using a linked Medicare-tumor registry database. *Medical Care*, **31**, 732–47.

Rao, C.R. (1973). *Linear Statistical Inference and its Applications*. 2nd edn. Wiley: New York.

Saunders, S.C. (1983). Statistical Estimation, Using Real Data from Systems Having a Decreasing Hazard Rate, and its Application to Reliability Improvement. *Reliability in the Acquisitions Process*. D. J. DePriest and R. L. Launer (Eds). Marcel Dekker: New York.

Shouki, M.M. and Pause, C.A. (1999). *Statistical Methods for Health Sciences*. CRC Press: London.

Smith, P.J. (1986). Estimation in linear regression with censored response. *Pacific Statistical Congress*, Eds. I.S. Francis, B.F.J. Manly and F.C. Lam. 261–5 Elsevier Science Publishers North Holland: Amsterdam.

Smith, P.J. (1988). Asymptotic properties of linear regression estimators under a fixed censorship model. *Australian Journal of Statistics*, **30**, 52–66.

Smith, P.J. (1995). On plotting renovated samples. *Biometrics*, **51**, 1147–51.

Smith, P.J. (1996). Renovating interval-censored responses. *Lifetime Data Analysis*, **2**, 1–11.

Smith, P.J. (1998). *Into Statistics*, 2nd ed. Springer-Verlag: Hong Kong.

Smith, P.J. and Peiris, L.W. (1999). Added variable plots for linear regression with censored data. *Communications in Statistics*, A, **28**, 1987–2000.

Smith, P.J. and Zhang, J. (1995). Renovated scatterplots for censored data. *Biometrika*, **82**, 447–52.

Stablein, D.M. and Koutrouvelis, I.A. (1985). A two-sample test sensitive to crossing hazards in uncensored and singly censored data. *Biometrics*, **41**, 643–52.

Therneau, T.M., Grambsch, P.M. and Fleming, T.R. (1990). Martingale-based residuals for survival models. *Biometrika*, **77**, 147–60.

Turnbull, B.W. (1974). Nonparametric estimation of a survivorship function with doubly censored data. *Journal of the American Statistical Association*, **69**, 169–73.

Turnbull, B.W. (1976). The empirical distribution function with arbitrarily grouped, censored and truncated data. *Journal of the Royal Statistical Society*, B, **38**, 290–5.

Wagner, S.S. and Altman, S.A. (1973). What time do the baboons come down from the trees? (An estimation problem). *Biometrics*, **29**, 623-35.

Weibull, W. (1951). A statistical distribution of wide applicability. *Journal of Applied Mechanics*, **18**, 293–97.

Wei, L.J., Lin, D.Y. and Weissfeld, L. (1989). Regression analysis of multivariate incomplete failure time data by modeling marginal distributions. *Journal of the American Statistical Association*, **84**, 1064–73.

Weissfeld, L.A. and Schneider, H. (1987). Inferences based on the Buckley-James procedure. *Communications in Statistics*, A, **16**, 177–87.

Wu, C.-S P. and Zubovic, Y. (1995). A large-scale Monte Carlo study of the Buckley-James estimator with censored data. *Journal of Statistical and Computational Simulation*, **51**, 97–119.

Zhang, C.-H. and Maguluri, G. (1994). Estimation in the mean residual life regression model. *Journal of the Royal Statistical Society*, B, **56**, 477–89.

Index

accelerated lifetime model 147
actuarial estimate 83
added variable plot 235
alive times 132
associated variables 44

baseline hazard 144
baseline survival function 148
bathtub-shaped hazard 30
Buckley-James method 196
   weights matrix 197

cdf 34
censor indicator 75
censor time 75
censored lifetime 73
censoring interval 204
change in fit 218
chi-squared distribution 128
coherent system 40
cohort lifetable method 80
constant hazard model 19
converge in probability 198
covariate 143
Cox conditional likelihood 170
Cox partial likelihood 170
Cox regression estimate 172
Cox-Snell residuals 157
cumulative distribution function
   method 48
cumulative hazard function 16,65
cuts 43

death 3
decreasing failure rate 30
degrees of freedom 128
deletion formulas 217
delta method 65
deviance 180

deviance residuals 180
DFR 30
distribution
   chi-squared 128
   Erlang 128
   exponential 19
   extreme value 27
   gamma 127
   Gompertz 26
   log-logistic 29
   lognormal 28
   Pareto 31,139
   Rayleigh 26
   Weibull 23
double-censoring 78
dual 44
dummy variable covariates 177

effective number at risk 83
efficient score test 176
empirical cumulative hazard function 66
empirical survivor function 5,55
empirical survivor plot 58
Erlang model 128
expected service life 33
explanatory variable 143
exponential 19
   hazard model 26
   probability model 20,127
   probability plot 157
extreme value probability model 27
Euler's constant 27

factor model 177
failure 3
failure time 3
Fisher information 120
fitted values 216
force of mortality 9

Index

gamma distribution 127
gamma function 23
Gompertz probability model 26
Greenwood's Formula 86

harmonic mean 49
hat matrix 204,216
hazard function 9
  baseline 144
  empirical cumulative 66
  cumulative 65
hazard model
  bathtub-shaped 30
  constant 19
  exponential 26
  power 22
  Rayleigh 26
hazard plot 67
hazard plot scores 67
hazard rate 9
HI values 216
hierarchical models 179

IFR 30
i.i.d. variables 5
increasing failure rate 30
indicator function 4
interval censoring 78,204
irrelevant component 40

Kaplan-Meier estimator 96
k out of n system 39

least squares
  estimator 201
  normal equations 188
  residuals 203
  weighted 203
left-censored 73
left-truncated 75
Lehmann alternatives 146
leverages 216
lifetable 81
  estimates 84
lifetest 80
lifetime 3
likelihood 119

Cox 170
likelihood equation 120
location parameter 27
log-likelihood 120,172
log-linear model 148
lognormal distribution 28,60
Log-rank test 176
lost to follow-up 73,81

Mantel-Haenszel Test 176
masking effect 241
maximum likelihood estimator 119
mean life expectancy 11
mean residual lifetime 10
mean time to failure 8
median descent time 78
method of scoring 120
minimal cuts 43
mismatch score 77
MTTF 8
modified Cox-Snell residuals 157

Nelson-Aalen estimator 108
Newton-Rhapson method of scoring 120
normal equations 188
number at risk 81,95

observed information 121
observed information matrix 134,170
order of the system 37
order statistics 79
outer product of vectors 228

parallel system 38
partial regression plot 235
partial residuals 189
paths 43
pdf 23
PL-estimator 95
plotting position 57
  for censored data 110
population quantile 60
population quantile function 60
power hazard model 22
power transformation 61
probability model
  chi-squared 128

Erlang 128
exponential 19
extreme value 27
gamma 127
Gompertz 26
log-logistic 29
lognormal 28
Pareto 31,139
Rayleigh 26
Weibull 23
probability plot 60
product law of reliability 42,48
product law of unreliability 42,48
product-limit estimator 95
proportional hazards model 144
proportional mean residual lifetime 17

quantile-quantile plot 60
QQ-plot 60

random censoring 125
ranked data set 55
Rayleigh probability model 26
redistribute-to-the-right algorithm 107
regressor variable 143
relevant component 40
reliability 37
   function 3
   ith component 41
   product law of 42
renovated
   dotplot 219
   hat matrix 216
   HI values 216
   leverages 216
   residuals 201
   responses 197
   scatterplot 197
renovation slope estimate 215
residual lifetime at age t 10
right-censored 73,75,95
right-truncation 75
risk set 95,168

sample percentile 59
sample quantile 60
sample quantile function 57
scale parameter 27

score function 120,172
self-consistent 103
series system 38
standardised mortality ratio 140
structure function 37
   increasing 40
support 4
survival function 3
survival random variable 3
survival time 3

total observed lifetime 129
total time on test 129
truncated data 74

Weibull
   index 36
   probability model 23
weighted least squares 203
weights matrix 197
withdrawals 73,81

zero-crossing 193